IMAGES
of America

LONG ISLAND'S
MILITARY HISTORY

The history of military activities on Long Island involves dozens of forts, hundreds of gun emplacements, and thousands of buildings. But, more importantly, it also includes the millions of men and women who served in America's armed forces with pride for over 200 years, stationed on the island or processed at its camps on the way to war or when returning home. These three tired, coffee-drinking veterans of the Spanish-American War, resting in front of their tent at Camp Wickof near Montauk in 1898, are typical of the civilians who joined so many others to form the military and naval units that protected America's freedom, often at the cost of their lives. (SC.)

IMAGES
of America

LONG ISLAND'S MILITARY HISTORY

Glen Williford and Leo Polaski

ARCADIA
PUBLISHING

Published by Arcadia Publishing
Charleston, South Carolina

Library of Congress Catalog Card Number: 2004104740

For all general information contact Arcadia Publishing at:
Telephone 843-853-2070
Fax 843-853-0044
E-mail sales@arcadiapublishing.com
For customer service and orders:
Toll-Free 1-888-313-2665

Visit us on the Internet at www.arcadiapublishing.com

Long Island's coastal defenses consisted of three separate groups of forts: one at the Narrows entrance to the city's Upper Bay, another at the western end of Long Island Sound at the entrance to the East River, and a third at the eastern end of the sound between it and Block Island Sound. The glacier that deposited its moraine as Long Island, along with the runoff from its melting, which opened passages in the moraine to form smaller islands, left ideal sites for forts, as this 1920 map of eastern Long Island indicates. (NA.)

CONTENTS

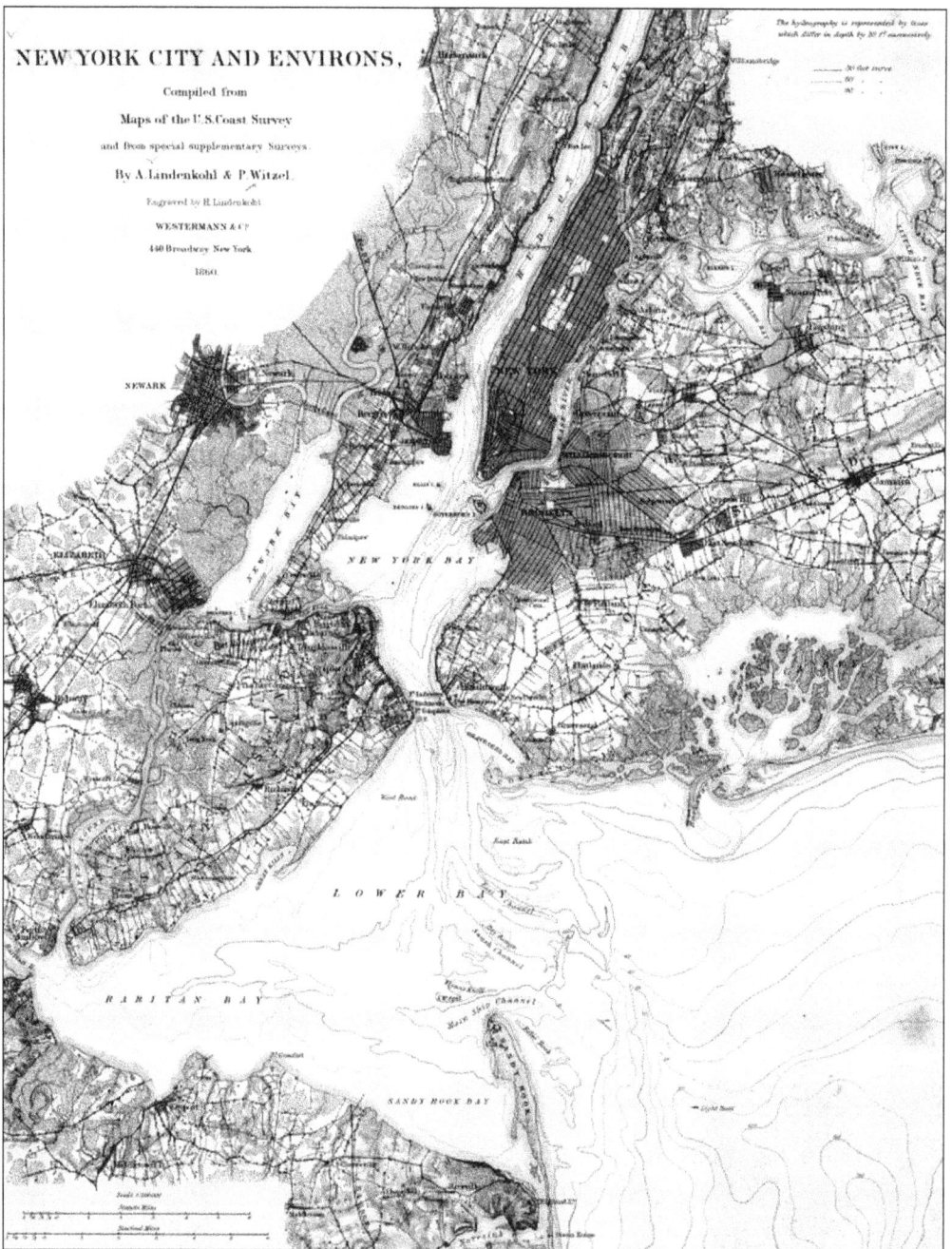

This 1860 engraving of the New York City area shows the Upper and Lower Bays, with the Narrows passage separating them, and New Jersey's Sandy Hook pointing toward this opening. From the 1600s, the Narrows was fortified, first by the Dutch, then the British, and finally by a series of American fortifications. Sandy Hook, the Navesink highlands behind it, and Rockaway Beach on the south coast of Brooklyn also received coastal forts to guard the main entrance to New York Harbor. (NA.)

INTRODUCTION

Long Island, New York, which stretches 100 miles north-northeastward of New York City and lies roughly 15 miles from the Connecticut shoreline, which it parallels, has had a long and varied military history because of its geography and topography. Created over 10,000 years ago by sand and boulders deposited by the last receding glacier, whose runoff formed deep channels that became the region's rivers, bays, and sounds and divided the island from the mainland, Long Island forms the eastern and seaward enclosure of New York City's great harbor, the "Gateway to the New World." Where its shores are near other headlands, fortifications were constructed to protect the harbor, its shipping, and its shipbuilding yards. Forts were also constructed where nearby smaller islands were left by glacial runoff, helping to close off Connecticut, New Jersey, and New York cities from possible enemy seaborne attack.

The last ice age also left grassy plains on Long Island, extending from Queens eastward into Suffolk County. Early aviators were attracted to this area because it offered a safe place to land wherever their short flights, fraught with frequent engine problems, took them. Soon, the possibility of cross-Atlantic flights tempted more aviators to this continuous open airfield, and many of them built planes for themselves and for sale to others. The nearness of New York City's financial backers, many of whom had homes along Long Island's "Gold Coast" and knew of aviation progress, caused entrepreneurs, aviators, and airplane builders to believe in the future of aviation on the island and to invest their money and skills there. It became the true "Cradle of Aviation."

The Port of New York and New Jersey's vast and sheltered harbor fostered the location of shipyards, ocean terminals, warehouses, and railheads along its miles of shoreline, and the bays and creeks of Brooklyn and Queens became sites for piers from which millions of servicemen and servicewomen and millions of tons of equipment and supplies left during both world wars. Shipyards, most especially the Brooklyn Navy Yard, found ideal locations along the island's shores, and submarine and torpedo factories chose Long Island's shallow bays for their activities.

The first defensive elements constructed on the island were forts: first of wood, earth, and brick; later of granite; then of concrete with earthen fronts; and finally of reinforced concrete covered overhead by earth. Army bases grew around them to house soldiers, and these locations were given the names of famous Americans, such as Alexander Hamilton, Gen. Joseph Totten, and Gov. Samuel Tilden, or of officers who served in the Civil or Spanish-American Wars, men such as Gen. Alfred Terry, Gen. Horatio Gates Wright, and Lt. Dennis Michie. The powerful gun batteries were designed by America's best West Point–educated engineers to be effective deterrents against naval incursions.

The defense of America's coastline evolved from total dependence on coastal fortifications, which were limited in range and vision as prospective enemies reached farther toward this country with their airplanes and undersea boats, to the use of airships and aircraft, which had the hundreds of miles of range and the aerial vantage points needed to fight these newer threats. Long Island was the ideal location for seaplane, dirigible, and blimp bases, and for conventional airfields from which reconnaissance planes, fighters, and bombers could cover vast expanses of ocean. Fields that were first just numbered soon gained the names of deceased aviators, and Roosevelt, Floyd Bennett, Hazelhurst, and Mitchel Fields quickly became known to local residents.

The men and women who operated all of these defensive elements, along with those who needed to be trained for overseas duty and for combat, had to be housed, fed, equipped, and readied; cantonments were established on Long Island for these purposes, starting in the 1860s with the Civil War. These camps, given soldiers' names, such as Scott, Black, Mills, Upton, and

Wyckoff, were constructed on the island's sandy terrain because this area was close to the port from which these soldiers would soon embark and because many of the men and women lived nearby when they were inducted.

Long Island factories produced America's two most successful fighter planes, some of its best navigational and naval gun-aiming equipment, its first guided missile, and the most powerful radar of its time, among many other defense items. The Grumman Aircraft Company's series of "Cats," among them the famous Hellcat that helped the U.S. Navy win its aircraft-carrier war against Japan in the 1940s, and Republic Aviation's "Jug," the rugged P-47 long-range fighter that helped defend Allied bombers during missions over Nazi Germany in World War II, helped to shorten the war.

With war plants come research laboratories and test facilities, and Long Island contained many. The first American aeromedical laboratory was begun during World War I in Mineola; Samuel Guggenheim's Sands Point estate held a navy development station; Sag Harbor and Montauk were the sites of torpedo research establishments; the army developed underwater mines at Whitestone; and both the army and the navy researched underwater listening devices on Fishers Island. The Sperry Gyroscope Company, Curtiss Aircraft Corporation, Grumman, Republic, and the Ranger Aircraft Engine Corporation, among others, also maintained extensive research and testing facilities at their plants.

Along with these manufacturing plants, most of which had their own employee schools, private aviation schools were opened on the island, many at Roosevelt Field. Thousands of pilots and mechanics learned skills that were directly useful to the armed services, and during World War II, Long Island's aviation schools trained military personnel in many needed technical specialties, just as they had trained civilians earlier.

With all of its tempting defense activities and its relative nearness to Europe, Long Island became a popular place for enemy spies to operate and, happily, to get caught. A German commercial radio station in Sayville transmitted coded sailing information about the *Lusitania*, resulting in its sinking by a submarine, causing many civilian deaths. The same station passed communications between Germany and Mexico regarding Mexico's joining in a war against America in exchange for several southwestern states. In 1917, the station operators were arrested and the station was seized. At Amagansett, four German saboteurs landed from a submarine in 1942, planning to destroy key industrial plants, but one informed the FBI about the others and about four more who had landed in Florida. Six were executed and two were imprisoned.

Although Long Island had an active and longstanding military presence, it was the site of only one battle. Weeks after the Declaration of Independence, Gen. George Washington led his militia army into Brooklyn from New York. They were pursued by the British, and a series of skirmishes and rear-guard actions in late August 1776 led to a secretive nighttime American withdrawal.

After more than 200 years, all of Long Island's camps, forts, air bases, and naval activities, and most of the industries that supplied military equipment, have closed or left the island. But what does remain, aside from parks, airports, and some structures, is the sure knowledge that the island's military presence once crucially aided great events in this country's history.

One

LONG ISLAND DEFENDS NEW YORK CITY

Fortifications on either side of the Narrows closed off this southern passage into the Upper Bay and its harbor. Staten Island, on the western side, was fortified from the 1600s through the 1940s with a series of increasingly longer-range guns. This 1932 aerial photograph shows Fort Richmond (begun in 1847) on the water's edge, Fort Tompkins (begun in 1858) on the bluff in the center, post–Civil War works (begun in 1871), and Endicott batteries (started in 1892) to the right of Tompkins. (GN.)

The primary weapon of the Endicott period of seacoast fortification construction, from 1886 to 1905, was the disappearing gun, a large 8-, 10-, or 12-inch rifled cannon mounted on a carriage that could be elevated to project the cannon above the concrete parapet for firing, as shown in this 1908 photograph of the rear of Fort Wadsworth's Battery Barry. The gun's recoil would then lower it to protect the gun and crew during reloading. To a seaward enemy, the gun would seem to disappear. (GN.)

Across the Narrows at Brooklyn was another fortified site. Fort Lafayette, the rectangular work standing offshore, began construction in 1812. It was followed by Fort Hamilton (onshore to the left of the pier), which began construction in 1825. A line of disappearing guns was added in front of the old fort early in the 20th century. The post ended its coast artillery role in 1948 but continues to serve as an active army reservation and as the site of a Veterans Administration hospital. (HM.)

The original masonry Fort Hamilton was built as part of the third system of American fortifications. These were funded as a result of a congressional initiative that provided a network of strong defenses for every important harbor. The work took from 1825 to 1831 to complete; when finished, Fort Hamilton held emplacements for 66 smoothbore cannons mounted in interior casemates and atop the ramparts. This 1860s drawing from *Gleason's Pictorial* is a view from the pier. (NA.)

Starting in 1890, the Fort Hamilton military reservation was extensively modified to provide emplacements for more modern and powerful disappearing guns. The Narrows side of the masonry fort was removed, and this series of gun emplacements was built immediately to the right of the old fort, as this 1959 photograph shows. The emplacements are empty, the guns having been removed some 10 years before. (HM.)

This surveyor is preparing for demolition that will remove Fort Hamilton's Endicott gun batteries in 1959 so that post housing can be erected along the Belt Parkway. Fort Lafayette, at left center, was completed in 1822. It served as a prison in the Civil War and as a magazine during both world wars, holding ammunition from ships entering the Brooklyn Navy Yard. It later became a storage facility for Hamilton but was removed for the 1964 construction of the Verrazano-Narrows Bridge. (HM.)

Because of their proximity to New York City's embarkation piers, Coast Artillery posts were also important as cantonments, at any time housing tens of thousands of troops heading to European combat and thousands of permanent party support personnel. This enthusiastic Women's Army Corps band is practicing on Fort Hamilton's parade ground in 1944. Bands marched in patriotic parades and played at troopship departures and arrivals. The dog is observing carefully, though he has probably heard this tune before. (HM.)

Recruits line up for chow at Hamilton during World War II. The uniforms of these E-1s are mixed because supply sergeants issued whatever stock and sizes they had on hand, which were sometimes leftovers from earlier quartermaster procurements. Six-compartment stainless steel trays were used in all services, but wartime shortages caused a switch to bakelite trays. The serving line and coffee urns were manned by KPs, transient soldiers assigned kitchen police duty for a day, and sometimes just grabbed out of line. (HM.)

This recruit's introduction to army life happened before his uniform was issued. He is a fireman, instructed by his 1942 Hamilton sergeant on stoking the coal-fired furnace so the barracks will stay warm throughout the night and there will be hot water in the morning. Up all night, he will carefully bank the fire, carry out the ashes, and, most importantly, build up the fire around 3:00 a.m. so it is roaring at 4:00 for reveille. Woe be to him if his buddies have to cold-shave. (HM.)

What appears to be a village street with a fort on the left is actually Fort Hamilton in 1897 with many civilian activities on post, not unusual in the 19th century. A hotel, photography studio, ice-cream parlor, and beer hall, among other things, line this post street. As transportation was difficult for soldiers, these businesses were welcomed, and they benefited from captive and restless customers who were paid regularly. (HM.)

The outermost of the southern defenses of New York's harbor was located on New Jersey's Sandy Hook peninsula. Its batteries, consisting primarily of 12- and 10-inch rifled guns, with two 16-inch guns added at Navesink during the 1940s, could interdict enemy vessels while they were out of range of the city. Both Fort Hancock, seen in this 1972 aerial photograph, and the army's first ordnance proving ground were located here beginning before the Civil War. (SH.)

A new post was established on Long Island's southern coast at Rockaway Beach to provide additional firepower on the ship channels in the lower bay. The first armaments mounted here were four six-inch barbette guns relocated from other coastal forts. Their emplacements were unprotected concrete bases with small adjoining magazines. Here, the first gun tube is being delivered to Tilden; the soldiers in this 1917 snapshot are the first soldiers to use it, albeit as a prop. (NA.)

Within a few years, Fort Tilden received its primary armament. Battery Harris was armed with two powerful 16-inch guns. These guns fired a 2,340-pound projectile over 25 miles to engage any warship on more than equal terms. In the 1940s, they were covered with protective concrete-and-earthen casemates. This battery, and that at Navesink, remained the most powerful elements in New York's harbor defenses until their scrapping after World War II. (GN.)

Soldiers are ramming two of six powder bags into the breech of one gun at Harris during a 1941 practice. The hydraulic chain rammer operated by the man standing on the right is similar to those used in naval turrets; it enabled the heavy shell and bags to be pushed snugly and quickly into the chamber. With the breech closed and the barrel elevated for firing, the breechblock recoiled into a pit beneath, where the men are standing. (GN.)

Fort Tilden's Battery Harris was modernized just before World War II. The gun's open turntable mounting allowed for all-around fire, but its concentric circles looked like the bull's-eye of a target from the air, which was perhaps too tempting for enemy bombers. This 1940s photograph shows one of the guns inside a newly constructed casemate made of tons of reinforced concrete and sand. A camouflage net was hung above the opening and draped over the barrel. (GN.)

16

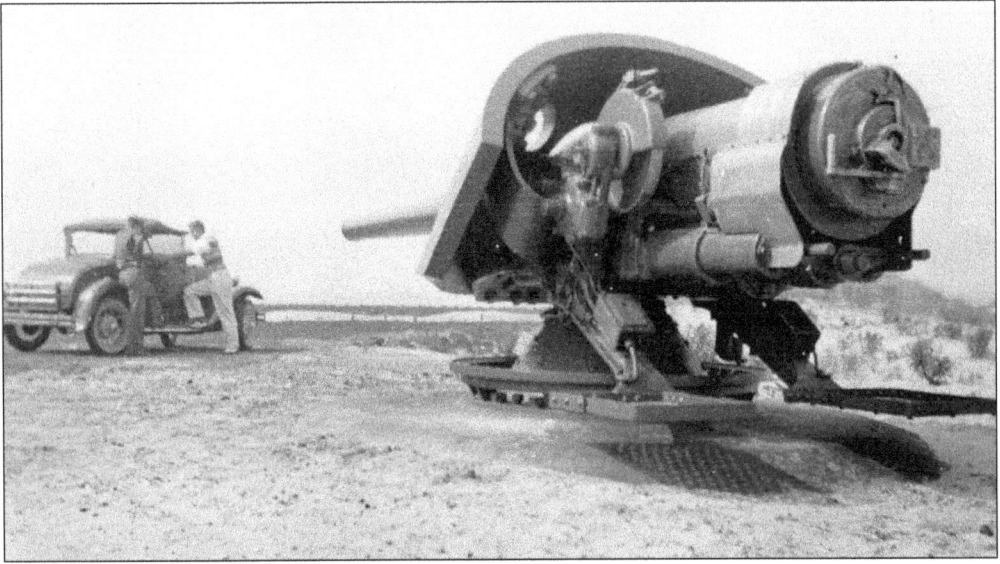

Not at all threatened by one of the open-mounted six-inch guns of Tilden's Battery Kessler, these visitors are enjoying the salt air in 1937. An enemy ship would not have been much troubled either, as even a nearby hit would disable the gun and injure its crew. During the war, one of the pairs of beach-mounted guns was rebuilt with additional overhead protection for its magazines, and an entirely new protected six-inch battery was added to the fort. (GN.)

Soldiers at Long Island's posts would be quite familiar with "hurry up and wait," the army's most often chosen method for gathering troops into formations. Early in 1941, these fellows of the 245th Coast Artillery, a New York National Guard unit stationed at Fort Tilden, are idly waiting outside the orderly room for their sergeant to emerge and begin yelling orders at them. A straggler is just leaving the barracks on the right, very happy indeed for the delay. (GN.)

Tilden was an important antiaircraft training post because it was the only area fort where live firing could be conducted. AA units had been assigned here since World War I, when the need to deter enemy planes became obvious. In 1928, three-inch mobile truck-drawn guns are practicing, and the ammunition railway for Battery Harris is being used to haul shells. The Coast Artillery Corps was chosen as the antiaircraft branch because of its expertise in firing at moving targets. (GN.)

The passage into the East River from Long Island Sound was fortified late because massive rocks at the aptly named Hell Gate limited navigation. Before the rocks were blasted in 1852, a masonry fort was completed in 1845 in the Bronx at Throgs Neck. Named Fort Schuyler, it was used by the army until 1934. In 1937, its 17 acres became the campus of the New York State Maritime Academy. This photograph, taken before a causeway was finished in 1961, shows the view looking northwest. (NA.)

In 1857, the army acquired land at Willets Point (directly across from Schuyler, seen in the background) and began a second fort to close the eastern passage. It was designed to be a multilayer, casemated water battery, but construction was halted during the Civil War and never resumed. This 1890s rear view shows the completed first layer of granite gun positions and the incomplete second tier. If finished, it would have mounted over 100 cannons. (TM.)

New gun battery construction soon modernized the defenses at Willets Point, renamed Fort Totten after engineer Gen. Joseph Totten. Work began on disappearing gun batteries in the early 1890s. A mortar battery, which delivered plunging fire on what were then lightly armored warship decks, was soon added, along with rapid-fire guns to deter fast-moving ships. This early-1900s easterly view shows the old fort and the earthen hill that hid and protected the new batteries. The painted sign warns boats not to anchor. (TM.)

Mortar batteries were also used in the Endicott System. Usually consisting of four pits of four mortars each, these 12-inch, high-angle, rifled weapons were designed to be fired simultaneously, rather like a gigantic shotgun, to have their shells land on vulnerable decks of enemy ships. Battery King, mounting 8 instead of 16 mortars, was built at Totten and paired with a four-pit battery at Fort Slocum near New Rochelle. Here, ready rounds for the next volley sit on their delivery carts. (TM.)

While soldiers formerly lived inside the dank casemates of early forts, Endicott Coast Artillery posts were cantonments with permanent barracks, mess halls, hospitals, service clubs, recreation buildings for their garrison troops, fine brick homes for officers' families, and guard houses for those men who became overly unruly and required a jail cell. This postcard from c. 1910 shows a Fort Totten enlisted men's barracks and a company of soldiers standing ready for morning inspection by their sword-carrying commander. (TM.)

20

The interiors of the barracks of a century ago were supposed to be kept as neat as this view from before breakfast until the duty day ended after supper. This Fort Totten dormitory, with its high ceilings, rifle racks, foot and wall lockers, neatly made beds, and polished floor, is typical of permanent posts. Privacy was never a consideration then for enlisted men, as they had to learn to be together continuously at a gun battery or in field combat. (TM.)

This photograph of a 1910 mess hall further shows the communal nature of military life. Before mess trays were introduced in the 1940s, thick china plates and coffee cups were used, and the scullery, beyond the hatch on the left, was always a greasy and steamy place. Both oil lamps and electric fixtures hang from the ceiling; municipal electricity was unreliable. Soldiers' comments concerning the taste of the chow and the ingredients tossed in by the cooks were the subjects of barracks humor then as much as today. (TM.)

The army's rapid expansion at the start of both world wars overwhelmed available military housing. Forts and camps on Long Island received hastily built, tarpaper-covered barracks until clapboard-sided buildings could be constructed. This Fort Totten barracks was retained throughout World War II, though, as it was still needed for troop housing. The windows are open to vent some of the heat produced inside a black structure on a sunny day. (CA.)

Employed for field training or for embarkation housing, tents were also necessary when a unit was assigned temporary duty away from its post. The soldiers of Fort Totten's 62nd Coast Artillery Regiment manned antiaircraft guns at Greenpoint, Brooklyn, early in 1942, and this is the main drag of their tent city. Duckboards helped with the inevitable mud, but the tents lacked floorboards, and earth was piled around their edges to lessen flooding. The soldiers must have missed their former tarpaper homes. (TM.)

22

Men of the 62nd, surrounded by Greenpoint's tenements, are practicing with one of their three-inch AA guns. Only aiming and loading could be rehearsed here, though live ammunition was kept ready should enemy bombers attack nearby fuel oil and illuminating gas tanks. Following Pearl Harbor, army units were dispersed to guard the island's coastline and key industrial facilities, though the need for antiaircraft protection lessened and this encampment closed in a year. (TM.)

Long Island's harbor forts, with their housing, hospitals, and piers, doubled as pre-embarkation sites, holding soldiers until troop ships were ready to sail from Brooklyn, Manhattan, or Bayonne. Men at Fort Totten are saying their goodbyes in the 1940s before boarding the *Ordnance*, operated by the army to carry cargo and men around the harbor. Steaming down the East River, it will dock at the Brooklyn Army Terminal, where the soldiers will march across the pier to their oceangoing transport. (TM.)

In the late 1890s, a series of new fortifications was built on the headlands and islands between Orient Point and the Rhode Island shore, protecting the passages between Block Island and Long Island Sounds. Five forts were established; three of them were modernized during the 1940s. One gun of the newly constructed Battery Barlow at Fort H. G. Wright, Fishers Island, which mounted two 10-inch guns, is seen c. 1910 from the rear as soldiers examine the raised weapon. (KS.)

Battery Dutton, one of three six-inch disappearing gun batteries completed at Fort H. G. Wright, is seen during a loading drill c. 1910. Though live-firing practices were conducted during summer training, they required clearing boats from a wide swath of Block Island Sound. Here, instead, soldiers are lowering the gun barrel by winching it downward after simulating firing; recoil would have lowered it if a shell had been fired. (NA.)

Two

CLOSING LONG
ISLAND SOUND

Dutton mounted two six-inch guns; here, a shell is being rammed into one's breach in 1902 while a soldier carries the bagged powder charge, ready to slide it behind. Other rounds are on the cart near the breech. The soldier on the platform is watching the towed target through a sighting telescope and moving the gun with a handwheel to follow the target. The range is set by elevating the gun, which is done before it is raised. (NA.)

Another 1902 photograph of Battery Dutton shows the gun in its elevated position, having just been fired. The soldier on the right, with the leather box of primers, has pulled a rope lanyard to ignite a primer he inserted in the breechblock, causing it to flame and ignite the powder bag in the chamber. Two other men, one an ordnance department inspector, cover their ears while another, on the sighting stand, is observing the shell's splash. (NA.)

In 1923, a serious accident occurred at Dutton. During National Guard firing exercises, the gun was accidentally discharged before it was fully elevated above the parapet. The inert round penetrated the concrete, and the gun recoiled off its elevating arms, its breech hitting the concrete loading platform. Two soldiers were killed; most of the others manning the gun were injured. The investigation concluded that the firing mechanism was defective, allowing it to be triggered even by the slight pull on the lanyard as the barrel rose. (NA.)

Twelve-inch mortars were emplaced at Fort Terry, on Plum Island, and at H. G. Wright. The eight mortars in the two pits at Wright were named Battery Clinton for the Continental Army general. After 1918, the armament was reduced to four mortars by removing the front guns in each pit. This postcard shows one mortar elevated to firing position, with soldiers practicing on the azimuth and elevation handwheels, while the other mortar has been lowered for a loading drill and to lean on. (LP.)

The most unusual weapon in Long Island's defenses was this 15-inch dynamite gun, erected in 1898 at Race Point, the western tip of Fishers Island. The gun used high-pressure compressed air to propel a dart-shaped shell loaded with TNT at enemy vessels without detonating the shell in the barrel. This "spitball-shooter" was never practical, and it was soon abandoned. Concrete remains still exist near Elizabeth Field, the island's former military airfield. (NA.)

Most large gun and mortar batteries had a plotting room where soldiers directed the accurate aiming of the weapons; this is Clinton's *c.* 1915. The setters on the curved bench have adjusted long brass arms on the plotting board to match the azimuth bearings telephoned from two distant observation stations, while the man with pillows under his chest is predicting where the shells and the targeted enemy ship should meet, as determined by angle-side-angle geometry, on a scale plan of Block Island Sound. (LP.)

When not practicing with the post's seacoast weapons, performing routine maintenance, pulling guard or KP duty, being inspected, parading, spreading rumors, hiding from sergeants, or scheming to get a weekend pass over to New London, soldiers at H. G. Wright lived in brick barracks such as this, now a grandstand for three-legged races. Field days and organized sports helped maintain morale, and officers knew that griping was not only normal, but a positive sign of the men's reliance on each other. (LP.)

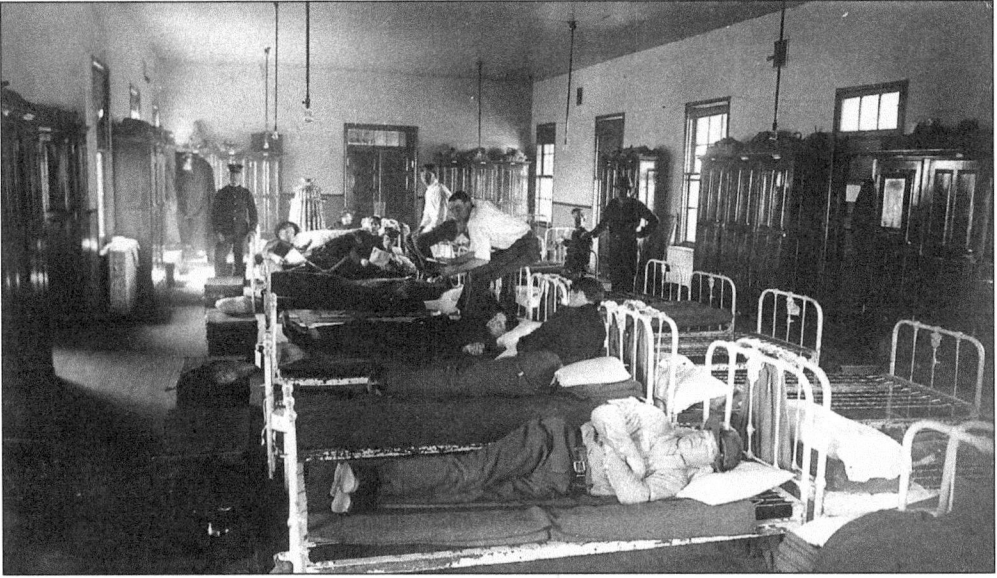

Peacetime units at posts such as H. G. Wright had their days filled with small, usually pleasureful events and lived as much like "civvy-street" as possible. Since their military purpose was to maintain the fort's equipment, to train visiting summertime units, and to become a cadre of skilled men prepared to lead an expand wartime army, there was plenty of downtime, despite efforts to create work to keep the men busy. Shooting the breeze, reading, and snoozing were quite popular in 1903. (KS.)

Hospital stewards were enlisted men with specialized jobs warranting better quarters. They assisted surgeons, bandaged wounds, managed hospital supplies and records, dispensed medicines, and supervised attendants. As did these two 1930s H. G. Wright stewards, they lived in apartments near the post hospital in quarters similar to those of the sergeants, separate from the troops. One man has served in many units, but only the 2nd C. A. Regiment was not at this post. Fort Ruger, Honolulu, though, was also good duty. (LP.)

The forts' orderly rooms were places of orders and orderlies; both are seen in this 1909 H. G. Wright office. Collections of troop rosters and daily orders are posted on the wall, and two company clerks who deal with such administrative details are ready for more. The Oliver typewriter, Model 3, is also ready. Such "visible writers," as well as carbon paper and mimeograph machines, allowed many more copies of formerly handwritten orders to be made and filed. (KS.)

Post quarters came in various sizes, and they were assigned by rank. While enlisted men lived in large barracks and married officers were given a substantial house, sergeants with families were assigned smaller wooden or brick cottages. This noncommissioned officers' (NCO) quarters at H. G. Wright, photographed with two fire-control observation stations in 1908, is typical of these single-bedroom shacks with a kitchen in the rear. Unmarried NCOs were billeted in single barracks rooms. (NA.)

Several states organized National Guard Coast Artillery units. These men held weekend drills at their local armories, equipped with dummy seacoast weapons and working fire-control equipment set up in their drill halls, and attended summer training for two weeks at seacoast forts. Here, New York's 13th Coast Artillery arrives at H. G. Wright in 1915. Live firings will be conducted with real guns at moving, ship-sized towed targets, a task that was difficult to accomplish back home. (KS.)

Summer camp was made as realistic as practical, and real sand went along with real guns. The comfortable brick barracks at H. G. Wright were for "Regular Army" soldiers, or RAs, and their summer guests had to make do under canvas. Long Island Sound's weather can be quite enjoyable in summer, with mild temperatures and ocean breezes, but a storm is approaching in the guise of their first sergeant. These fellows appear ready, though, for his inspection of their uniforms and humble quarters. (KS.)

Long Island Sound forts were hosts to several significant army and joint army-navy maneuvers in the early 1900s because warships could operate here without impeding commercial shipping as they would at the main channels into New York's harbor. These soldiers in the 1909 exercises are constructing a temporary gun emplacement at an H. G. Wright beachfront using the most plentiful material at hand and employing the most easily available digging and transport device. (LP.)

The second island to be fortified was Plum Island, situated off Orient Point. All of its 797 acres were acquired by 1901, and construction of most batteries was completed by 1905. Named for Gen. Alfred Terry, who served in the Civil War, Fort Terry's largest guns were at the eastern end to protect passages into the sound. This photograph, showing two soldiers and the cantonment prior to World War I, was taken at the island's center, looking west. (KS.)

View of Fort Terry, N. Y.

Officers' quarters, built in the early 1900s, once lined the north side of Terry's parade field. Most were of brick, but the earliest were of wood-frame construction. Majors and colonels lived in single homes, while most company-grade officers' quarters were duplexes, holding two families. Unmarried officers had apartments in bachelor officers' quarters, the BOQ. A large post would have six batteries of artillerymen, requiring homes for some 40 officers. All of the homes have been demolished. (KS.)

The handsome quarters at Terry, which gave a sense of the importance of commissioned officers, were supplemented with less impressive buildings 40 years later. By January 1941, the serious business of expanding a peacetime army in preparation for war produced a different environment for all ranks. These are unpainted wooden officers' and enlisted men's quarters east of the main post, remote from its amenities. Their style speaks of the coming emergency. (NA.)

Construction of adequate post housing always lagged behind needs, especially between wars, when Congress cut military appropriations and many believed a standing army was unnecessary. In 1900, this sergeant stands in front of his jerry-built tarpaper shack at Fort Terry. Because he was an enlisted artillery or engineer specialist, his choice was to live away from his family or bring them to the island, live in this hovel, and wait to receive better quarters when another man retired or was transferred. (NA.)

Given the shortcomings of early-1900s post construction, it is obvious why the fire brigade should be well trained and equipped. Even brick barracks were built with wooden floors and beams and used oil lamps and coal stoves. Smoking was also permitted, though not when reclining on a bed. The Fort Terry fire detachment, having pulled its hook-and-ladder wagon to this NCO quarters in the 1920s, is practicing evacuation and roof venting of the wooden structure. Their pumper is out of view. (NA.)

Organized sports were an integral part of the activities at coastal forts. Games provided diversion and exercise and fostered teamwork and physical fitness. Further, teams from other posts could be challenged, allowing for pride-building rivalry and, some would say, heavy betting. Nine members of the 88th Coast Artillery's 1913 baseball team pose at Fort Terry, and they appear fairly well equipped. Profits from the post exchange supplied the men's recreational funds. (MH.)

INSPECTION. FORT TERRY, N.Y.

Of course, close-order drill and company inspections and parades also built morale and spirit, or so the troops were told. Spacious reservations on the larger islands, such as H. G. Wright and Terry, contained ample parade fields, the grassy centerpieces of their cantonments, with barracks, officers' quarters, post headquarters, a bandstand, a very tall flagpole, and the retreat gun arranged around them. These 1905 soldiers at Terry are standing for inspection by their sword-carrying lieutenants. (MH.)

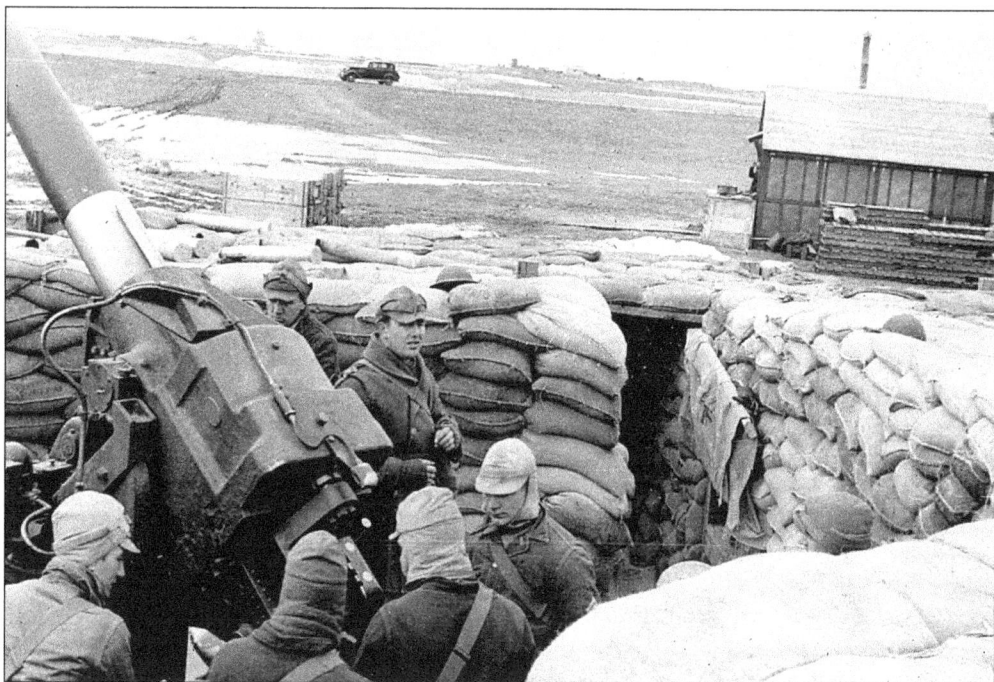

As the prospect of another war approached, the most critical activity at Long Island's forts became training on their defensive weapons, especially the relatively modern antiaircraft guns. On a 1930s winter day, gunners in a dugout emplacement engage in target practice using one of Fort Terry's three-inch AA guns. These batteries had been emplaced on the islands in the early 1920s to provided air defense for their seacoast guns; during the war, they were replaced with more modern 90-millimeter batteries. (JP.)

Great pride is shown here, the result of the gunners' good aiming. This canvas target sleeve was towed over Terry for these men to shoot at, and one of them marks a hole with his finger. A shell did not have to actually hit a plane to bring it down, as it exploded at a height predetermined by the time set on its fuse and its shrapnel would, hopefully, strike the plane. Later, radio proximity fuses improved accuracy. It is hard to tell if this hole would have been effective. (JP.)

36

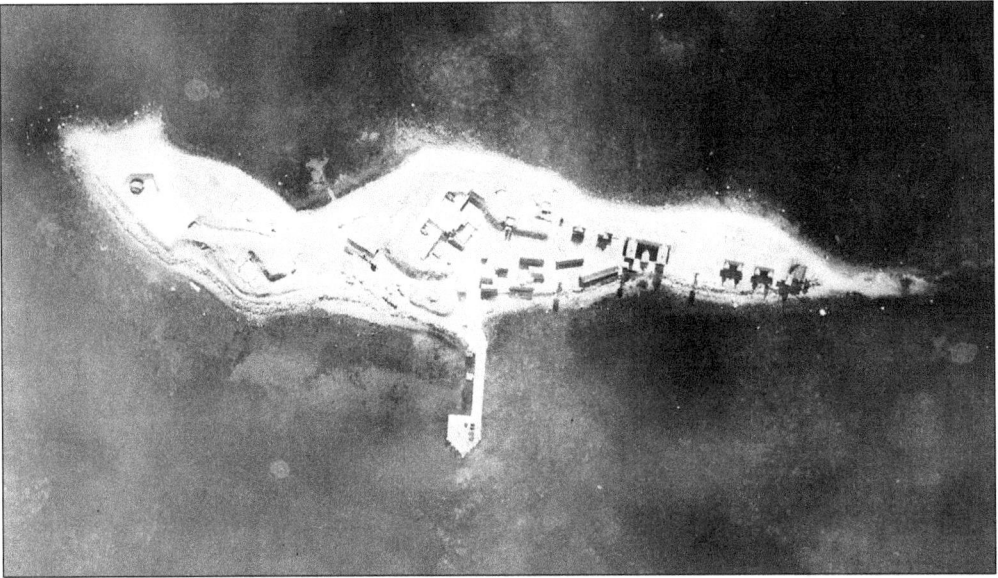

Fort Michie was established in 1896 on 19-acre Great Gull Island. This 1924 aerial photograph shows the fort; its major gun batteries, seen at the northern tip, on the left, and at the middle, all face toward the Race, the main passage into Long Island Sound. A large company barracks, two officers' quarters, and the post hospital are on the right, with warehouses and NCO housing near the wharf. (NA.)

Battery Palmer, in the center of Great Gull, mounted two 12-inch disappearing guns. It was named for Col. Innis Palmer of Civil War and Mexican War service, and it was one of the most important works of the eastern defenses, effectively covering the Race with two of the largest guns available. This 2002 aerial view also shows the three-inch guns of Battery Pasco on the north shore and a control building for underwater mines, which was built at the rear of Palmer for protection. (GW.)

One of Palmer's guns is seen on its disappearing carriage in the 1930s, its breech wrapped to protect it from salt spray. It is in the lowered loading position, and its great lead counterweight is raised in a pit below the carriage, ready to be tripped to bring the gun up for firing. Aimers' platforms are on either side of the barrel, and there is an optical sight bracket on the far side. The wide loading platform was needed to handle a 40-foot rammer. (JP.)

Buddies posing by their guns was a favorite subject for snapshots, which certainly impressed the folks at home, including girlfriends. These men and their mascots are ranged on one of Palmer's weapons in the 1930s, and one has rotated and partially opened the breechblock. When the guns were loaded, a wheeled cart carrying the 1,046-pound projectile was rammed at the run against the open breech, starting the shell into the chamber as soldiers pushed against it with a ramrod. (JP.)

Some six-inch guns were mounted not on disappearing carriages but on pedestals, allowing them to be traversed rapidly by hand to follow fast enemy destroyers and minesweepers. Michie's Battery Maitland carried two of these weapons, whose crews were protected by a heavy steel shield rather than a high concrete parapet. Crews in the 1930s had painted "Block" and "Montauk" on the wall to help them orient the rapid-fire gun when given aiming information by the battery's optical rangefinder. (JP.)

The largest gun in Long Island Sound's defenses until the creation of long-range batteries at Fishers Island and Montauk Point in the 1940s was this unique 16-inch disappearing gun at Fort Michie. Its massive concrete pit was completed at the eastern tip in 1922; space being limited, a 10-inch battery was demolished to make room. The shield protected the crew from concrete shards if an enemy shell hit the parapet and from dirt and gravel tossed around by the muzzle blast. (NA.)

Battery J. M. K. Davis could fire a 2,340-pound shell 22 miles once each minute. Its range was less than the gun's potential because the carriage could not elevate the barrel above 30 degrees. The Race was protected, though, from an enemy run-by, an attempt to steam past a fort before its guns could cause damage. The battery's shortcomings were the open pit, which was vulnerable to aircraft, and the shock to the carriage caused by firing and by the 610,000-pound counterweight slung below it. (JP.)

Manually loading a 16-inch gun was dangerous. A standard-gauge railroad brought shells from the wharf into the battery's magazines for storage. There, chain hoists moving on an overhead rail brought the projectiles to heavily constructed shot carts, which carried them to the loading platform. A shell would then be shoved into the breech, as is being done in this photograph. The crew manning the ramrod is ready, and carts with further shells are lined up outside the magazine corridor. (MH.)

An accurate method of obtaining direction and range information was determined in the 1900s. Directional sightings taken at the same time from two widely spaced telescopes were telephoned to a plotting room, where the target's position was found by triangulation. Azimuth and elevation settings were then calculated and telephoned to men at the guns. Low-lying Michie required higher observation stations, and tall but obvious towers were built c. 1910. The farthest contains the plotting room in the lower level. (MH.)

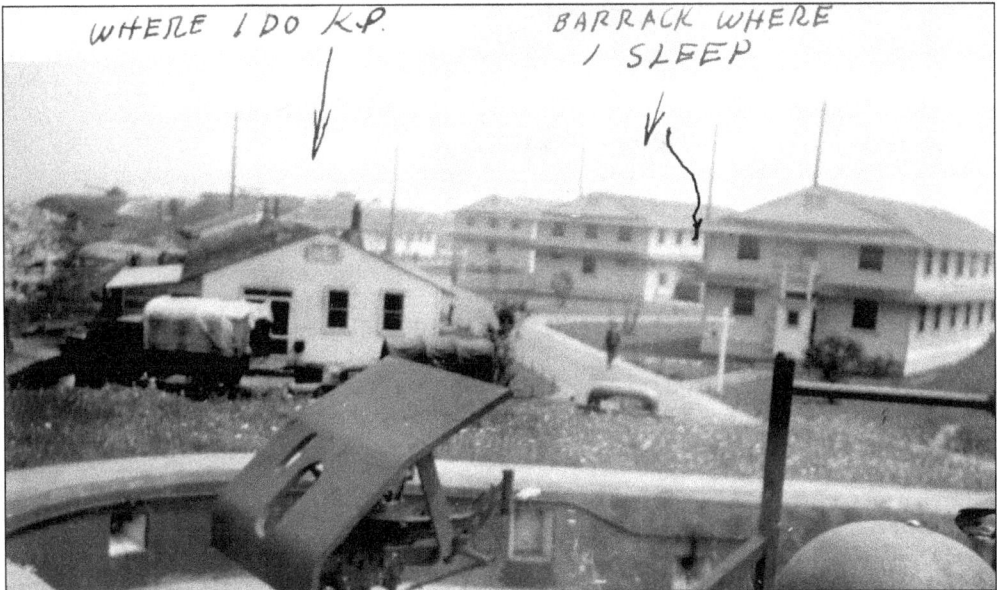

This 1940s snapshot was annotated by a soldier to show his family what he considered to be the most important aspects of his life at Michie. Taken from Battery Pasco, which mounted two three-inch rapid-fire pedestal guns, it shows some of the wooden barracks and support buildings recently constructed to house the additional soldiers stationed there during the war. The barrel has been removed from this gun for repair or replacement, leaving the shield strangely empty. (JP.)

41

"Dress right, dress!" was just bellowed at these Fort Michie soldiers in the 1940s, words familiar to every soldier ordered to "fall in." These men are not recruits but part of the post's complement of regular artillerymen. They carry gas masks under their left arms and wear World War I–style leggings and ankle-high trench boots, which were replaced during World War II with higher combat boots that better supported the ankle. (JP.)

One of Michie's soldiers, standing his four-hour guard duty with his unloaded rifle near Battery H's barracks, poses in this snapshot for a friend. On an isolated island, one might wonder about the purpose of having a guard on duty 24 hours a day, every day. But military life was disciplined, ordered, and traditional, and men had to be kept occupied on post and reminded that they were still soldiers. A guard also watched for fires, a constant danger with wooden buildings and stiff breezes. (JP.)

Soldiers in this late-1930s snapshot are busily cleaning the .30-caliber rifle ammunition stored in these wooden boxes, polishing the brass cartridges to prevent salty moisture from corroding them. Going AWOL from their island was hardly a choice for these men. Movies, listening to the radio, reading, having a beer in the evenings, and sports also kept the men occupied, though hitting too many home runs could soon end the game. (JP.)

This antiaircraft spotting station was built atop Michie's unused Battery Davis. In poor weather, a soldier could watch from inside the glass cupola, warmed by a coal stove, but this sunny 1940s day brings two soldiers outside. Before World War II's radar, observers with binoculars were the most reliable way of detecting enemy aircraft, so one hopes that the fellow focusing his on the buddy who is taking this snapshot will soon look skyward. The more studious one is probably not looking at airplane silhouettes. (JP.)

A seasoned "cosmoliner" is instructing new men in the proper use of one of Fort Michie's three-inch antiaircraft guns, mounted high on Battery Palmer's flank so as to fire over the cantonment areas. Cosmoline is a heavy grease preservative with a distinctive and lingering odor, and ordnance mechanics and coast artillerymen were ribbed about how they smelled. Three-inch guns did not have the vertical range to cope with long-range bombers, but neither could German bombers cope with such a wide ocean. (JP.)

This day has been chosen for AA practice at Michie in the late 1930s. Cases of three-inch shells have been brought from underground magazines, and the ends of the sealed tubes have been removed. Sandbags around the gun protect its crew from shrapnel and machine gunfire, though they are not immune from naval gunfire. While two of the men are out of uniform, their informality was usual before the war, and commanders seldom pressed the point on a hot summer day. (JP.)

44

These World War II soldiers are spending a hot 1940s day on a patch of land at Fort Michie, some playing horseshoes and others shooting baskets at a backstop that doubles as a clothesline pole. One looks longingly at the Connecticut shoreline; just 4 miles away, it might as well be 4,000 miles to a man stationed here for months. Despite the insects, the screens have been removed from the barracks windows to allow more air to circulate. (JP.)

Nothing tastes quite as good as cold beer at the end of a hot day. Michie had its post exchange, but these 1940s men have chosen their company dayroom to gather and pose for this snapshot. To lessen drunkenness, the army limited the alcoholic content of beer consumed on its bases to 1.4 percent until the 1940s, when it became 3.2 percent. Officers could purchase hard liquor and regular beer; they did their social drinking in more refined settings, but they could not have had more fun. (JP.)

Fort Mansfield was established on the Rhode Island shoreline at Napatree Point, a spit of land west of Watch Hill, to guard the channel between it and the eastern tip of Fishers Island. The construction of three batteries was accomplished between 1898 and 1902, and a cantonment area built toward Watch Hill along the beach. Here, soldiers climb from the magazine level to the gun platforms of Battery Wooster before World War I. All batteries had attached enameled porcelain naming signs. (CM.)

Battery Wooster, named for Connecticut Revolutionary War hero Gen. David Wooster, was armed with two eight-inch disappearing guns, one of which is seen here pointed down the fort's only street. The carriage has been rotated to its extreme left azimuth; further movement would be stopped by the concrete parapet. Mansfield's other armament consisted of four five-inch guns in two batteries. (OM.)

Essentially a sandbar, Napatree Point, while a good location for armament, turned out to be a poor place to build a fort. Storms and beach erosion ate away the sand, and Mansfield's guns were removed for use as field artillery in World War I. The fort is shown here c. 1918. The buildings washed away in the hurricane of 1938, when 121-mile-per-hour winds and 18-foot storm surges submerged the point. Wooster's concrete survives; a five-inch battery is now out to sea. (LP.)

Fort Tyler was built on a spit of land at the northern end of Gardiners Island. Its design was unique. A concrete box with an open area in the middle, it was shaped like older granite forts to provide all-around defense. Completed in 1898, the post was never armed, as the need to protect Gardiners Bay and Sag Harbor diminished and the army discovered that its reservation was getting smaller. This 2003 aerial view shows Gardiners Point, which was once connected by a roadway to its namesake island. (GW.)

This 1905 photograph shows the inside of Tyler, with an emplacement for one of its two eight-inch disappearing guns seen on the right. Two five-inch pedestal guns were also to have been mounted inside, firing over the concrete parapet. Storms undercut the walls and washed away the roadway, providing Prohibition rum runners with an isolated boat shelter. The Ruins, as fishermen know Tyler, was used by the navy as a target for bombing practice during World War II. (CM.)

During the 1930s, Long Island Sound's defenses were modernized. The entrance to Block Island Sound was fortified with two massive dual 16-inch batteries at Montauk Point; a similar battery was built at Wilderness Point on Fishers Island, and three more were located at the entrance to Narragansett Bay. This July 1942 photograph shows the construction of a gun pit and the entry into a 500-foot-long gallery, which contained magazines and a power room and connected with the other casemate. (NA.)

The emplacements at Montauk's Camp Hero included the two dual 16-inch batteries and a dual 6-inch battery. The 16-inch guns, which fired a shell 25 miles over the ocean or onto Block Island, should an enemy land there, were protected with thick reinforced concrete with earth piled atop to deflect shells. Four-inch steel shields and an overhead canopy protected the gun house openings. Only proof-fired by 1944, one of the guns is shown then, with two men nearby. (LC.)

Wyandannee Inn, Montauk Point, Long Island

When the 650 acres at Montauk Point were acquired in the late 1930s, existing private structures were purchased with the land. Some were demolished, but those that had potential military use were retained. The 1920s Wyandannee Inn, seen in a contemporary postcard, was used as housing for the crew manning the nearby six-inch battery, and they made good use of its decidedly non-military bedrooms, lounge, and dining hall. Some officers were billeted in retained summer bungalows. (LP.)

The 1940s long-range batteries required widespread observation stations for accurate triangulation of their distant targets. At dozens of locations along Long Island's coastline, on Block and Fishers Islands, and as far north as Point Judith, land was purchased and new stations were built. Often disguised to resemble typical seaside cottages, many still exist and have been converted into residences, though all have a thick concrete tower inside. The narrow observation slit in this Hither Hills house reveals its origins. (GW.)

The army also had a navy, and some of its ships were mineplanters, such as the *Colonel John Story*, seen here in New York's harbor in the 1940s. Underwater minefields were planted to deter enemy ships and submarines and to destroy any that entered a protected harbor. Observation stations tracked approaching ships, and mines could be set to go off on contact or could be detonated electrically when a ship passed nearby. All mining equipment was kept on shore, tested and ready for loading aboard ship. (GN.)

50

Mines loaded with 3,000 pounds of TNT are arranged on a mineplanter's deck, soon to be swung overboard by one of the ship's cranes at predetermined positions. Used in shallow channels, these mines were set on the bottom, but in deeper locations smaller mines were buoyed to 1,000-pound anchors and allowed to float below the surface, unseen but ready to sink an enemy ship. Because planting required skilled teamwork, frequent drills with dummy mines were conducted. (GN.)

Eighteen Nike missile batteries guarded the New York City area from the 1950s through the 1970s; six of them were located on Long Island. This aerial view shows Rocky Point's launch site; its radar and control area was a half-mile away. Sixteen liquid-fueled missiles were stored underground and raised to the surface for launching. Here, two are elevated on the far launch area, while others are serviced nearby. Earthen berms surround each facility, limiting the damage that would be caused if a missile were to explode on the ground. (CA.)

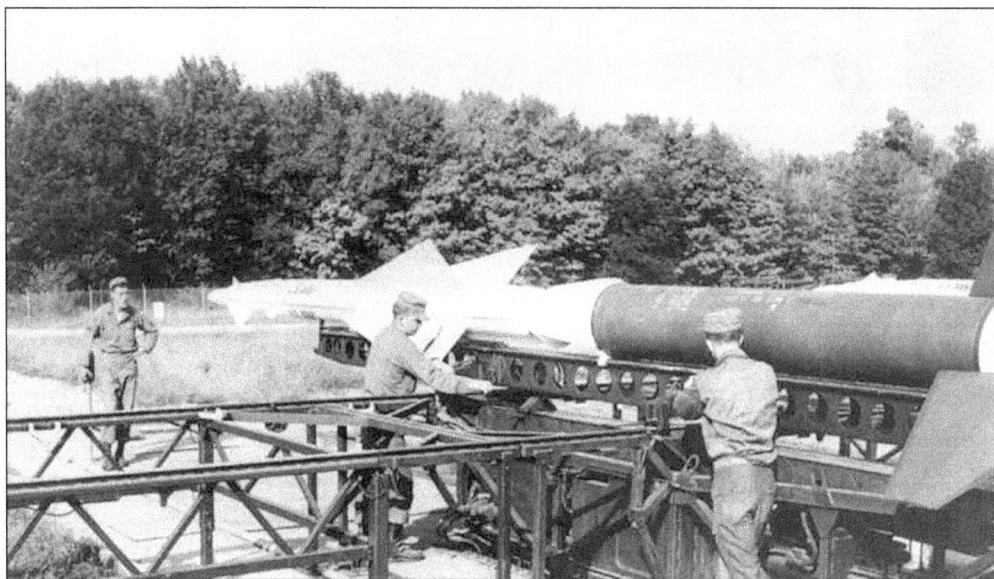

After being brought above ground one at a time on an elevator, missiles were pushed along twin rails and positioned above an electrically operated elevating beam for raising and launching. These 1960s soldiers are practicing that process, ensuring that the carriage holding the 2,400-pound missile rolls smoothly. Nike Ajax missiles had a range of just 25 miles, but the later Hercules version traveled over 85 miles and could be fitted with an atomic warhead; none of Long Island's sites contained these warheads. (CA.)

Soldiers inside a launch van are watching a radar screen and automatic plotters, which are tracing the paths of a simulated launched missile and its target. In this drill, frequently held at every site, the paths should intercept, indicating a destroyed aircraft. Three radars were employed: an acquisition search set and missile and target tracking radars. Computers preset the missile's gyroscopic navigational system, launched it, and guided it in flight. Yearly firings were done at Fort Bliss, Texas. (CA.)

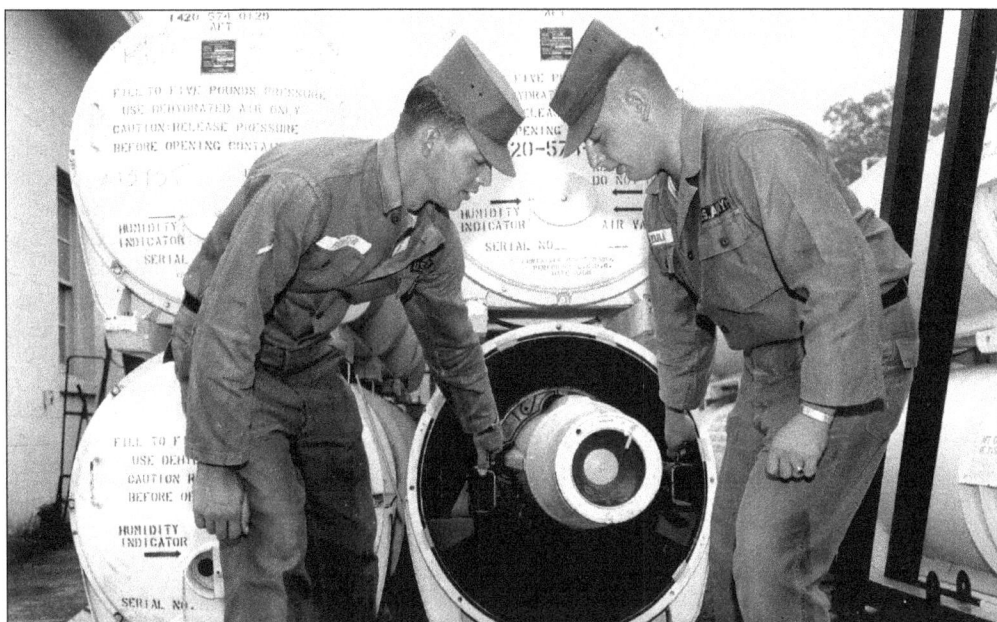

Long Island's Nike bases were supplied with electronic and mechanical repair parts, as well as with spare missiles, from Bellmore's Army Supply Depot. In the 1960s, two soldiers inspect missile engines stored in pressurized and humidity-controlled cylinders. These are not fueled and could be trucked to any launching site that required a replacement engine. The triple warheads were stored at the sites in bunkers. (HM.)

Four BOMARC (Boing Michigan Aeronautical Research Center) surface-to-air missiles are raised from their Westhampton launching buildings for testing in the 1960s. With a range of 250 miles and a ceiling of 60,000 feet, they were intended to destroy enemy bomber formations by exploding a nuclear warhead in their midst. The site was operational from 1959 to 1964. Rocket boosters for this missile were unstable, fueled with corrosive nitric acid, alcohol, and kerosene, and a 1960 BOMARC explosion in New Jersey released several ounces of plutonium. (CA.)

The most powerful radar of the 1960s was mounted at Camp Hero to detect enemy planes approaching New York City. This Sperry AN/FPS-35, whose 150-foot antenna rotates atop a five-story windowless concrete transmitter building, has a range of 200 miles. Near it, two smaller antennas rotate inside inflated rubber domes while an airliner, flying unusually low, passes in the distance. U.S. Air Force personnel were housed in the old Coast Artillery buildings, built to resemble a seacoast fishing village from the air. (CA.)

Inside Hero's operations building, Federal Aviation Administration (FAA) air-traffic controllers track civilian planes; air force personnel nearby track military aircraft. Planes carried radio transponders that identified them on the screen by a distinctive number. When a controller touched the plane's spot on the screen with a probe, that information was sent to a nationwide system of SAGE (Semi-Automatic Ground Environment) computers that allowed adjacent radar sites to track the aircraft's flight. Interceptors would be scrambled to identify unknown planes. (CA.)

Three

THE PORT OF
NEW YORK

The New York Navy Yard, established in 1800 and known locally as the Brooklyn Navy Yard, was one of the first federal construction shipyards. The site chosen was 200 acres of swampy land on the East River at Wallabout Bay, once the location of infamous Revolutionary War British prison ships on which 11,000 captured men died from starvation and disease. Years of development at the yard are evident in this 1939 aerial view. (NA.)

T. F. Rowland was one of Brooklyn's builders of iron ships in the 1860s. His firm, Continental Iron Works, located near Newtown Creek, was responsible for the hull construction of a radical new turreted warship designed by John Ericsson. Eventually commissioned USS *Monitor*, it was armed and completed at the nearby Brooklyn Navy Yard. This "cheesebox on a raft" became famous when it fought the ironclad CSS *Virginia*, formerly the wooden USS *Merrimac*, at Hampton Roads, Virginia, in 1862.

Crewmen of the *Monitor* are shown on deck, reading, playing a board game, and smoking. The interior of the cramped steam-powered, iron-hulled ship was never pleasant, especially on a summer afternoon, and it is easy to understand why these men prefer the open air. A cookstove was also brought on deck, and sailors prepared meals and ate there. The dented rotating armored turret, fitted at the navy yard with two 11-inch Dahlgren guns, is behind them. (LC.)

Dry docks and building slips were needed to construct large naval warships, and the Brooklyn Navy Yard was well equipped with these. America's first dry dock was completed here in 1851, and six others were added later. This is the large 1918 double slip and its overhead cranes, built to launch ships needed during World War I. Scaffolding has been erected at both shipways in preparation for hull construction, and the gantry cranes are ready to carry hull plates. (NA.)

In 1940, navy yards began two-shift schedules, increasing their production in anticipation of war. The Brooklyn Navy Yard was then constructing two battleships: the 45,000-ton *Iowa* and the 35,000-ton *North Carolina*. This worker is operating a vertical boring machine, milling out the hollow center of a cylinder to be used in the machinery of one of these warships. The yards employed highly skilled workers and were equipped with vast machine shops, the best in the world for naval construction. (LP.)

Ships completed at the Brooklyn Navy Yard include some of America's most famous warships. Best known are the *Arizona*, sunk on December 7, 1941, at Pearl Harbor, when Japanese aircraft surprised the fleet by attacking before their country declared war; the *Missouri*, aboard which the Japanese surrender took place at Tokyo Bay in 1945; and, seen here *c.* 1896, the *Maine*, lost through an internal explosion in Havana in 1898, an event that helped precipitate the Spanish-American War. (NH.)

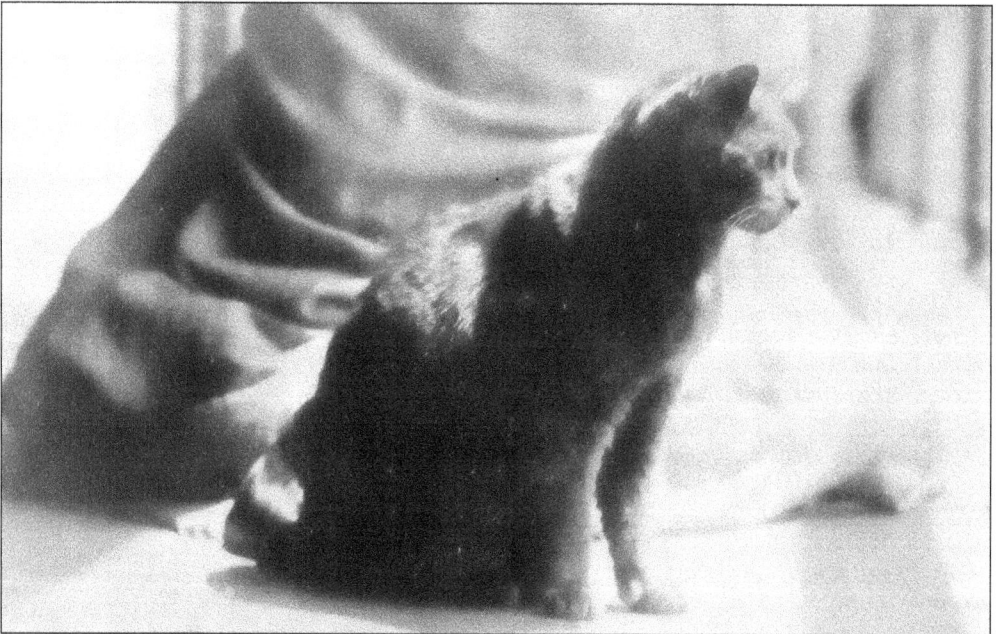

The explosion of the *Maine*'s forward magazine, probably the result of spontaneous combustion of coal in an adjacent bunker, caused the death of 266 sailors. Only 88 sailors survived, along with this lucky fellow, found clinging to a mast the next day by rescuers searching for dead and injured men. Most naval ships of the 19th century carried mascots, and cats were favored. After his experience, he is destined for shore duty rather than reassignment to another vessel. (NH.)

58

In 1925, the submarine *S-51* was accidentally rammed by the steamship *City of Rome* while it was on the surface at night in Block Island Sound. Only 3 of its 36-man crew survived. By pumping air into undamaged compartments and attaching pontoons, the boat was raised in 1926 and towed to the navy yard. Here, dry dock No. 4 is being dewatered as navy investigators examine the vessel. Although much of its machinery was operable, its keel and hull were beyond repair, and it was sold for scrap. (LP.)

The monitor *Amphitrite* was launched in 1864. Despite its low freeboard, it steamed to Cuba in the Spanish-American War. Seen here in 1918, it was a guard ship in New York's harbor, challenging entering ships and protecting the navy yard. Coaling was the worst task in the navy, as dirty coal sacks had to be carried below, and dust covered every surface of the ship for days. Here, sailors are digging lumps of coal out of the barge, which is filled with broken coal and dust. (KP.)

A Curtiss HS flying boat, with instruments designed by Lawrence Sperry and manufactured at his father's Long Island City plant, rests at the yard in the 1920s, having been hauled out of the East River after testing. This was a pusher plane, with the propeller behind the engine. The gyroscopic unit that operated a leveling stabilizer is under the front hood, and the plane's Sperry compass is mounted before the control wheels. Sperry also developed a radio-controlled airplane in Amityville, to be used as a guided missile. (CA.)

A Loening OL-3 observation amphibian rests on a catapult mounted atop a gun turret, probably of the USS *Pennsylvania*, which was docked at the yard in the 1920s. Turrets were used because they could be rotated to launch planes into the wind. The propellant that moved the sled carrying the plane along the rails was a powder charge ignited by the gunnery division, which tried to fire as the ship rolled up, lest the plane be shot downward into the water. (CA.)

Extensive railheads and supply depots were needed to embark and support an American Expeditionary Force "over there" in France during World War I. Several Atlantic coast bases were constructed, including one at Bayonne, New Jersey. The largest was the Army Supply Base, later known as the Military Ocean Terminal, located in Bay Ridge, Brooklyn. Two massive warehouses, two dozen rail sidings, and berths for 20 oceangoing ships were completed in 1919. Finished too late for this war, the base was well used in the next. (NA.)

The army used the Bush Terminal Company buildings as its first Long Island troop embarkation base. This Bay Ridge property was leased in 1917, and its seven piers and huge warehouses quickly became a busy military facility. Street railway lines connected it to the nearby Army Supply Base, and the navy added adjacent warehouses for its use. Its complex of wharves and storehouses, extending inland to Second Avenue, can be seen in this 1920s aerial. (NA.)

Camp Black was established on the Hempstead Plains in March 1898 to muster troops for the Spanish-American War. The first unit to occupy it was the 71st New York Regiment, seen here setting up tents as a private guards rolls of canvas and piles of tent poles. The encampment was short lived, but this site was later chosen for the First World War's Camp Mills, named for Gen. Albert Mills, who won the Congressional Medal of Honor in Cuba in 1898. (SC.)

The 33rd Michigan Infantry leaves an East River ferry in Brooklyn to march to the railroad train taking them out to Camp Black. Its location was chosen because it was next to the Long Island Rail Road, allowing troops and supplies to be easily transported. Bedrolls are slung over the left shoulder and rifles over the right, and one of the more fastidious soldiers is carrying a hatbox for his Stetson campaign hat. (SC.)

Four

TRAINING FOR WAR

The only American camp established specifically to detain soldiers returning from a foreign war was located east of the village of Montauk. Camp Wikoff, named for Col. Charles Wikoff, the first officer killed in Cuba, was used to quarantine and treat soldiers who had possibly contracted malaria, dysentery, and yellow fever in the Caribbean. Its best known unit was Col. Theodore Roosevelt's 1st U.S. Volunteer Cavalry. Here, the future president is inspecting some of his pet dogs, just arrived from Oyster Bay. (NA.)

Troops from Wikoff are parading in Southampton in September 1898, just before they were mustered out, and townspeople have arrived on foot and in carriages to honor them. Some 29,500 men went through Wikoff, and about 350 died there from their illnesses. Theodore Roosevelt's Rough Riders honored Long Island's only president with a statue, Frederick Remington's *Bronco Buster*, which is now at Roosevelt's Sagamore Hill home, a unit of the National Park Service. (SC.)

Tents of the 42nd Rainbow Division dot a field at Camp Mills, Mineola, in 1917, needed because the tarpaper-covered temporary buildings had not been finished. The division's nickname came from Maj. Douglas MacArthur's remark that this National Guard division "would stretch over the whole country like a rainbow" because it was comprised of men from many states instead of one, as was the custom. One of its regiments was New York's "Fighting 69th," immortalized by a movie featuring Pat O'Brien as Chaplain Francis Duffy. The 42nd entered combat in France in 1918. (MH.)

This 1919 view shows some of the temporary wooden buildings at Camp Mills. They are most likely part of the 2,000-bed hospital, judging by the overhead steam pipe and the enclosed connecting corridors from the rear of the barracks on this street. Mills was built to house some 40,000 pre-embarkation troops, in addition to 5,500 garrisoned there permanently and 500 more predicted to get themselves into the camp stockade. It also held returning troops. The camp closed at the end of March 1920. (CA.)

Half of Mills's transient soldiers lived in tents before they went "over there," and open mess halls and field kitchens were knocked together to feed them. While outdoor dining might be pleasant during Long Island's hot summers, and insects were kept at bay with screens, fly paper, and the constant use of swatters, more substantial buildings would seem necessary in a few months. Instead, because the war was expected to end soon, tents were erected for cooking and eating. (CA.)

The danger of fire was present at every cantonment, all of which contained wooden buildings, flammable canvas tents, coal stoves, and soldiers who smoked. Fire brigades were established and manned by the station complement, non-transient soldiers who operated camp facilities. This 1918 postcard shows one of Mills's firehouses and the proud firemen assigned there to the chief's car and pumper. The "U.S.Q.M.C." on its hood indicates that the Quartermaster Corps was responsible for operating this equipment. (LP.)

Yaphank's Camp Upton, which was begun in 1917, was Long Island's largest base; it could house and train 43,000 infantrymen. This photograph shows construction along 25th Avenue, with railroad tracks at 72nd Street. The streets were named for those in New York City, and, yes, there was a Broadway and a Times Square. This line of barracks will soon be filled with anxious trainees. Closed after the war, the camp was rebuilt in World War II as a reception center, convalescent hospital, and prisoner-of-war camp. (NA.)

A decidedly non-military formation resulted when these recruits were called out on their first day at Camp Upton. Looking around and waving at the camera will be discouraged in the time-proven army way, which can be properly explained to them only by a drill sergeant. Not just their behavior and clothing will change; France will profoundly alter their lives. All will mature; some will not return home, while others will be sent to Europe again just 22 years later. (GW.)

Group photographs were frequently taken at training camps by local civilian photographers and sold to the men as remembrances of the experiences they shared. These soldiers of the 77th Division's 305th Infantry Regiment, stationed at Camp Upton, are posing outside the mess kitchen attached to their barracks late in 1917. Because of labor problems, road and barracks construction had not been completed, and soldiers worked as laborers to finish the camp. (LP.)

These Upton trainees, led by the officer on the left, are undergoing the time-proven regimen of calisthenics, though it seems as if two of them are still learning to keep in time with the commands. Their next exercise will be 20 pushups, which will improve both their muscles and their paying attention. The unpainted barracks on the left are typical of the hasty cantonment construction of World War I; they were not intended to last and were built cheaply, as the separate latrine building between them indicates. (LP.)

It is difficult to know what the photographer said to these men of Company J, 306th Infantry Regiment, to cause them to pose this way, but he probably caught them on their way to the chow hall, as some do appear impatient. Food was an important health and morale concern, and a company commander was responsible for ensuring clean mess facilities and properly cooked food. Medical department staff inspected kitchens and tested a camp's water, as illnesses caused as many casualties as combat wounds did. (LP.)

Training at Upton was made as realistic as practical. Not only combat skills but also field encampment procedures had to be learned. This field bakery is being set up even before a tent is stretched over the wooden frame erected to protect the sacks of flour. Civilian laborers are unloading more sacks, while the soldier who drove the truck takes a catnap. One quickly learns in the army that "sack out" has nothing to do with lugging heavy flour bags. (LP.)

These recruits are ensuring that they will sleep well at Upton. Cotton-stuffed mattresses were not available, and the men are stuffing their own mattress bags with straw, obtained from the bales dumped next to their barracks. One problem was that a soldier never slept alone on a straw mattress; lice were his constant bedfellows and had to picked off in the morning. Because of disease and the danger of fire, straw was later replaced with cotton wool. (NA.)

Camp Upton   Base Hospital Ambulance   122.

Cantonments needed infirmaries, hospitals, and ambulances. This is Upton's World War I crew in front of the base hospital. One horrible day of their basic training cycle would find a company of soldiers lined up at the door of an infirmary, sleeves rolled up, to receive typhoid inoculations. These painful injections were needed to prevent this disease caused by contaminated food or water, a real possibility in France's muddy, unsanitary trenches. Hopefully, this would be the worst shot any of the men would experience. (GW.)

Unlike Mills, Upton was vast enough to permit live firing practice and realistic instruction using tanks, hand grenades, and simulated gas attacks. While calisthenics and KP built discipline and fitness, days in the rifle pits were a vivid reminder to soldiers of why they were in the army. On a cold day in 1917, these men are firing their Springfield bolt-action rifles at distant targets, while cadre next to them instruct them on proper handling. Others behind them keep a tally and try to keep warm. (LP.)

70

World War I armies still relied largely on horses and mules for transportation. The animals required greater care than motorized vehicles did, but they started reliably. The thousands of horses required by a division needed proper feeding, veterinary care, and rest, and they were as vulnerable as men to combat injury and death. Some horses used for training purposes at Upton are fed and housed in this tent stable. (NA.)

Proudly showing off his three most prized military possessions, this World War I soldier sits on the wooden sidewalk outside his barracks. His Springfield is intended to kill enemy soldiers, but his boots may kill his feet. Nicknamed "little tanks" by the doughboys, these trench boots have hobnails clinched in their soles and metal horseshoes rimming their heels. They will not wear out, but they also will not flex when he walks. Waterproofing was accomplished by applying tallow and neats-foot oil. (LP.)

A World War I fire at Camp Upton threatens to spread to tank wagons being loaded with oil from railroad tank cars. Since the tracks pass near the cantonment area, some wooden barracks could be engulfed, though wide firebreaks between groups of them should contain the fire. The wagons were pulled by steam-powered tractors, which were probably the source of ignition. To keep down summertime dust from unpaved roads, oil was sprayed on them, a practice that was also common on civilian roads. (LP.)

Because of the combat experience their armies had already gained before America entered World War I in 1917, instructors from France and Britain worked alongside U.S. Army officers at Upton to train the Yanks whose assistance they so badly needed. Fortunately, many Long Island civilian airplane pilots quickly enlisted at Mitchel Field, were commissioned, and became flight instructors to the thousands of young men eager to join as fighter pilots. These are some of Mitchel's shavetail lieutenant instructors. (CA.)

72

Training maneuvers gave soldiers practice under field conditions and allowed their commanders to judge the effectiveness of tactics and new equipment. The army fought the navy in Block Island Sound in 1902, when Gen. Arthur MacArthur sank all four navy battleships while Adm. Francis Higginson was simultaneously destroying all gun emplacements at H. G. Wright. Despite such optimistically reported successes, these men studying their Gatling gun seem unconcerned, and their corporal on the right seems mighty proud of them. (NA.)

One technology evaluated at the maneuvers was searchlights, needed to illuminate attacking ships at night. This Sperry 60-inch light, with a range of four miles, has been mounted atop a wooden tower erected on Fishers Island by soldiers who, despite their carpentry skills, have also anchored it against prevailing strong winds with guy wires. Most coastal forts were equipped with searchlights, some mounted on rail cars or elevating metal towers to conceal them before use. (JP.)

Before similar firing ranges at Upton were completed, Long Island's soldiers went for summer training upstate at Plattsburg, together with men from other regions. These infantrymen are pulling targets on one of the post's rifle ranges. The other half of their platoon is 1,000 yards away firing M1903 Springfield rifles, and these men are marking the location of hits they make with metal discs on long poles. "Maggie's drawers," a red cloth, is waved when the target is completely missed. (LP.)

The men of Fort Totten's 62nd Coast Artillery antiaircraft regiment are spending a lazy 1926 summer day camped out in Mineola as part of their field training. Their bivouac area is the air base's grassy edge, where they can set up tents and their unit's three-inch and .50-caliber weapons while they watch planes take off and land. These enlisted soldiers are in two-man "pup" tents, shelter halves that connect to form a small sleeping (and afternoon napping) tent. (TM.)

Long Island soldiers were also sent to Pine Camp, near West Point, for training in the years before World War II. These men are cleaning their mess kits using hot water produced in clean garbage cans by immersion heaters, underwater stoves that boiled the water in half an hour. The first, where the men are lined up, contains soapy water, and the next two contain rinse water. Food grease left in mess kits produced a quick case of the "GIs," causing excessive toilet paper use. (LP.)

Even before 1942, when black airmen began combat training at Alabama's Tuskegee Army Air Field, some Harlem residents were qualifying as pilots at Roosevelt Field in a quasi-military unit, the Harlem Air Squadron. Their first inspection in uniform is being held at the field in 1935. One woman member is on the left. In the 1930s, blacks also gained their coveted pilot licenses at Howard University through the nationwide Civilian Pilots Training Program. (CA.)

One of an inductee's first experiences in World War II was taking the Army General Classification Test, given at a reception center such as Upton, where he will spend a week before being assigned to a basic training base. He was tested in vocabulary, arithmetic, spatial relationships, and inferring missing information to predict which army jobs he was best qualified for. Manpower requirements often meant, though, that men had to be sent to infantry schools no matter how their tests turned out. (LP.)

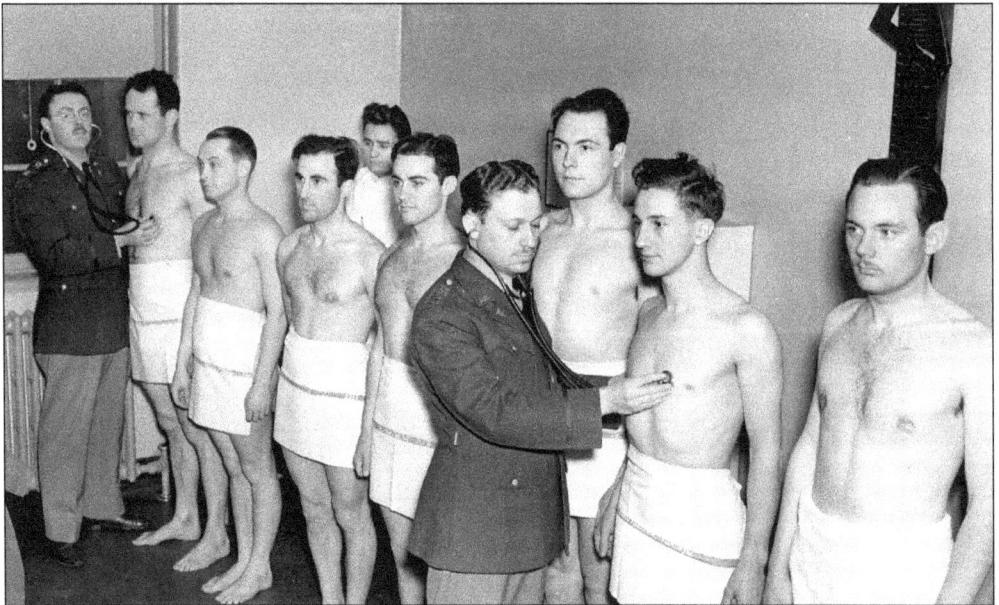

Flight school cadets line up for their medical examination at Mitchel Field in 1940. This is the easy part of their physical; touching their toes while being examined for prostate enlargement will soon follow. Before World War II, such exams could be leisurely, but soon mass screenings became commonplace, and the towels and the two doctors checking heartbeats were eliminated. Medical standards for pilots were more rigorous than they were for other specialties; men who did not pass served the Air Corps on the ground. (CA.)

This hapless private is about to ride Mitchel Field's Barany Chair in 1940. The chair was spun and twisted sideways, backwards, and forwards, evaluating the man's reactions to flight. The medical orderly is testing his blood pressure before the trip. America's first aeromedical research laboratory was begun at Hazelhurst in 1918 and was transferred to Mitchel in the 1920s. This chair was also used with the Ocker Box, into which a subject placed his face to test disorientation. (CA.)

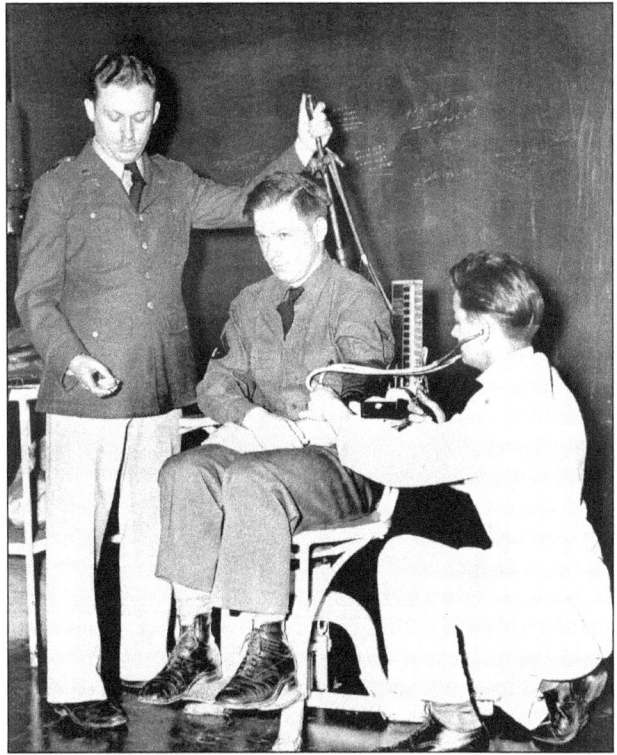

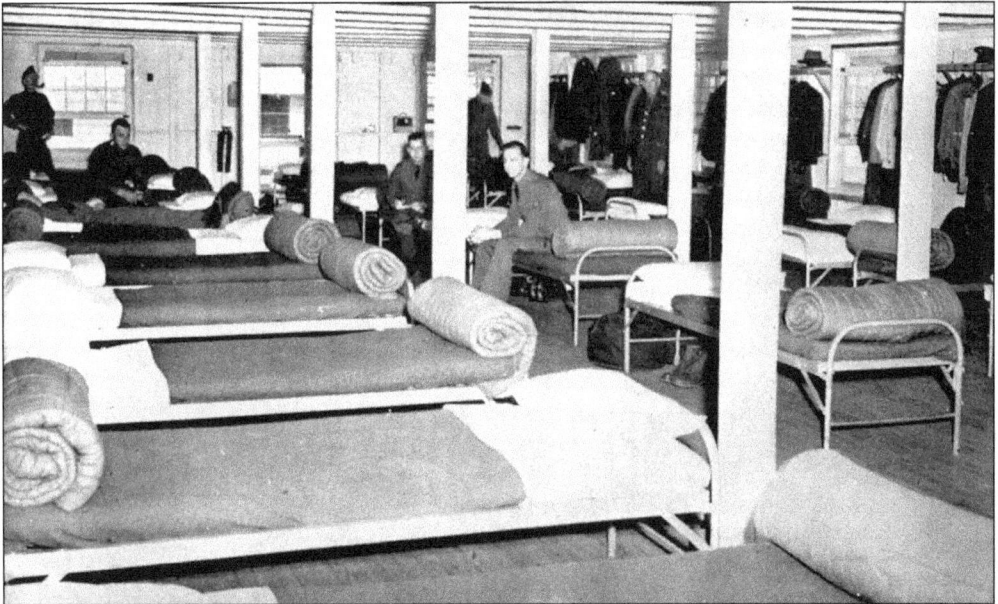

World War II men sit in a typical reception-center barracks awaiting the next order to march to somewhere unknown for another of the army's seemingly endless and mindless routines. They are being prepared for a soldier's regimented life, which includes obeying orders without debate or hesitation. Since the heads of the beds along the walls in this crowded barracks were to have been alternated, one of them is soon to have a special, and long-lasting, lesson in listening to a sergeant's commands. (LP.)

A soldier's wartime life gave him few opportunities to sit around. These men at Fort Totten in 1941 are on "police call," army terminology for cleaning up their cantonment area. New buildings have not yet been painted, but that will soon be done nationwide as a morale booster. Concrete or asphalt will replace wooden sidewalks, and streets will be paved. The box outside the furnace-room doors is a coal bin, and the one behind the barracks is for garbage cans. (TM.)

These Wacs, Women's Army Corps personnel, worked in medical, technical, and clerical positions at Mitchel Field. Here, they have completed training in the post's gas chamber. Inside the wooden structure, the women were required to stand for minutes with their masks on while tear gas was released, proving that the masks would protect them. In basic training, soldiers had to remove their masks in gas-filled rooms and chant their names and serial numbers while breathing some gas. When they were finally allowed to leave, none walked out. (CA.)

78

Floyd Bennett Field in Brooklyn was the city's first municipal airport, but it was located too far from Manhattan. When La Guardia Airport was opened in the late 1930s, the navy and the Coast Guard took over Floyd Bennett for training and operations. Its handsome terminal building and control tower is shown in the 1930s; it became the navy's headquarters and operations facility. Passengers walked across the apron to board planes, but on rainy days they could go through tunnels. Here, the exit hatch of one of the tunnels is open. (CA.)

Training included live gunnery practice, conducted over the ocean at towed targets, but these Marine Corps aviators are demonstrating techniques on the ground to a group of students at Floyd Bennett. The gunnery sergeant instructor, with his .30-caliber Lewis gun mounted on a Scarff ring in the rear observer's cockpit of a Curtiss F8C-4, may well be mentioning that shooting directly to the rear will quickly remove the plane's tail assembly, resulting in a "kill" for the enemy. His wide belt secures him while he is standing during unsteady maneuvers. (CA.)

Realistic training in fuselage repair was provided to these navy mechanics at Floyd Bennett when a Grumman XF2F-1 flipped over on landing after a flight from Anacostia Naval Air Station near Washington, D.C., in 1937. Acceptance tests were accomplished at Anacostia for these carrier-based fighters, nicknamed "Flying Barrels," which were constructed with metal fuselages and fabric-covered wings. Squadrons of them were aboard the *Lexington, Ranger, Saratoga,* and *Yorktown* before World War II. Bent propellers were not unusual, and airfields and carriers stocked many spares. (CA.)

These Women Appointed for Volunteer Emergency Service (Waves) recruits are listening to a 1951 demonstration at Floyd Bennett. Slated to close and become a unit of the National Park Service in a few years, the base was still being used as a navy school after the war. The members of the Waves branch of the U.S. Navy are being familiarized with all aspects of shore duty. Some will serve as parachute riggers, while others will work in communications, supply, and medical departments. (TM.)

This derelict plane at Floyd Bennett is being used to teach firefighting techniques to the crash crew. A ladder against the wing permitted the downed pilot, a heavy dummy made of rope, to be rescued, and the men are moving in on the burning plane with foam from fire engines off to the right. Practices gave the crew members confidence that they could walk into the fire of a crashed aircraft, wearing flame-resistant hoods and turnout gear, to rescue an injured pilot. (CA.)

Members of a wartime flight crew are being briefed before takeoff in a Quonset hut at Floyd Bennett Field. They flew Vought OS2U Kingfishers and Grumman J4F Widgeon amphibious observation planes, searching for German submarines and for sailors whose ships had been torpedoed. Their flight path has been sketched on the blackboard, and a reminder about the differences between American and enemy subs is on the right. The Kingfishers carried machine guns, and the Widgeons carried bombs; both would attack a suspicious submarine. (CA.)

Though they were not established to prepare students for wartime service, in the 1930s Long Island's many aviation schools trained mechanics who would all too soon have to use their specialized skills in military service. Young men at the Roosevelt Aviation School are practicing the skill of doping an aircraft's fabric wing using cellulose acetate or cellulose nitrate to shrink and tighten the linen covering. Dope also strengthened and waterproofed the fabric, made tight to maintain the wing's airfoil shape. Eleven coats were brushed or sprayed on. (CA.)

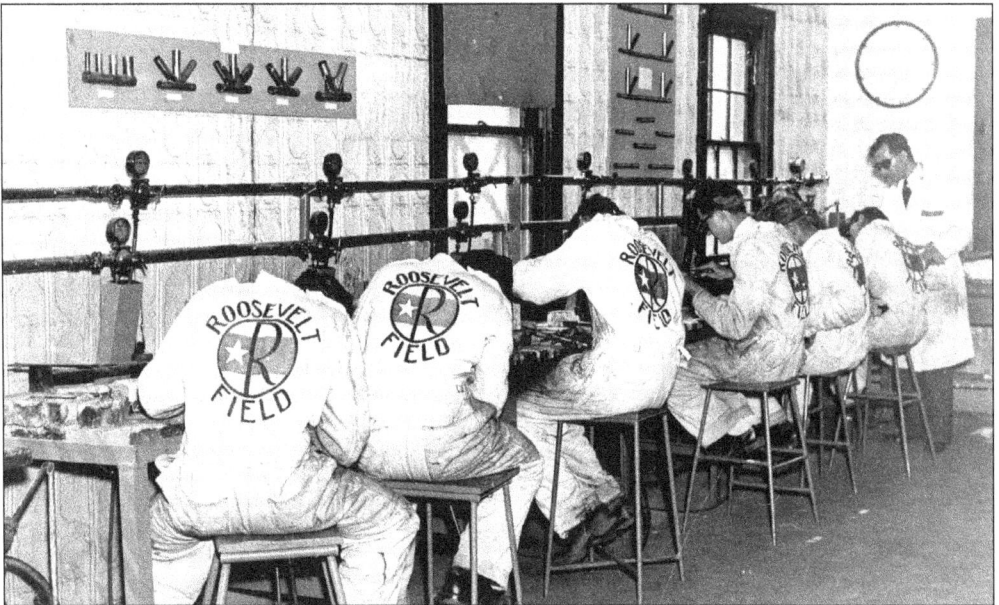

While early biplanes were constructed with wooden frames, by the 1930s most training planes used stainless steel tube fuselage frames that were joined by welding. Fighters and transports were turning to aluminum frames and skin. Students at the Roosevelt Aviation School are bent over a stone bench, practicing the crucial skill of oxy-acetylene welding while an instructor guides them. Samples of proper welds are displayed on the tin wall, and all the men wear dark glasses to protect their eyes from the flame's glare. (CA.)

These civilians at Jericho's Faust Aviation School are using air-powered rivet guns to rivet a plane's fuselage skin in the 1930s. Another student inside the plane bucked the rivet as the gun shaped its head, clamping the thin aluminum to the plane's frame. Many thousands of practice rivets were correctly set before a student would qualify as an airframe mechanic, and his first project was a sheet-metal toolbox, an object of pride and practicality. Every correctly formed rivet was needed for pilot safety in combat dogfights. (CA.)

The Engine Air Service school at Roosevelt Field trained mechanics before and during the war to overhaul complicated radial power plants. Military and commercial engines required disassembly, measurement of wear, and replacement of rings and worn parts about every 300 hours. These men are learning to use specialized tools supplied by engine manufacturers; experienced repairmen will also teach them techniques to ensure that the engines will not fail in flight. Pistons and cylinders removed during overhaul are on the shelves. (CA.)

Edwin Link patented his "Training Device for Student Aviators and Entertainment Apparatus" in 1929. He developed the device based on his pilot's license training and on the vacuum-operated player piano technology he learned while working in his father's organ company. The trainer permitted students to learn many airborne skills on the ground, reducing training crashes and cost, but was seen first in penny arcades and traveling carnivals. At the Roosevelt Flying School, this 1930s student will lower the hood to practice "blind" or instrument flying. (CA.)

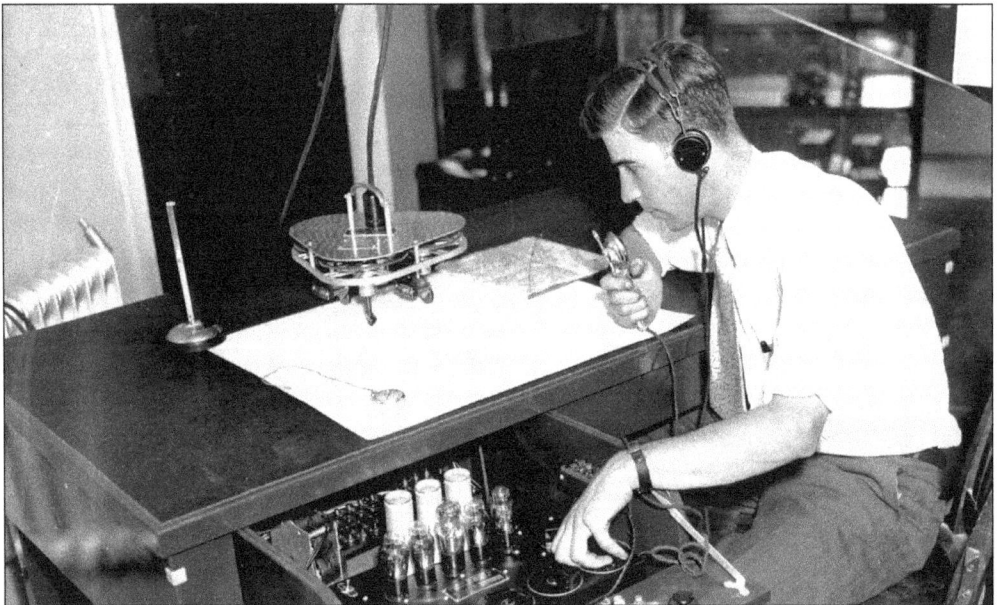

An instructor is watching his course traced on a map by the moveable "crab" that responds to the simulated flight. The student can bank, pitch, and turn the trainer with the same controls found in a real cockpit, and he can also crash, miss the runway, get lost, or run out of fuel—with only hurt pride. With this control panel, the instructor introduces mechanical problems, produces rough weather, and communicates with the student. World War II armed forces used over 10,000 Link Trainers. (CA.)

Five

THE HEMPSTEAD PLAINS

Hazelhurst Field or (Field #1)

Mineola's Hazelhurst Field, whose grass landing area is in the center of this 1917 photograph, was named for Lt. Leighton Hazelhurst, killed in an air crash. One of the pilots trained there was Lt. Quentin Roosevelt, the former president's son, who died in combat over France in 1918. The eastern part of the field was renamed for him. Wooden barracks and mess halls line dirt roads at the top of the photograph, while hangars along Old Country Road on the right mark the northern edge of the base. (CA.)

Most pilots trained in a Curtiss JN-4 biplane, affectionately known as a "Jenny." Many were sold after the war to barnstormers, usually former military pilots who flew them in air shows and made extra money by taking passengers, hopefully pretty girls, aloft for their first airplane rides. Springtime mud was a problem on the airfield and around the barracks, even though roads were ditched and drained. A plane could easily flip over if a wheel ran into a soft patch of ground. (CA.)

Pilot trainees were not the only people who were busy at Hazelhurst during the war. The base was also a repair and supply depot for other Long Island fields, and its shops, such as that seen behind the International truck, were busy patching up the planes damaged in the inevitable daily training accidents. The drivers of this army truck are fortunate to be out of the mud for now, and the truck has a chain on the rear wheel to assist its travel through the road's goo. (CA.)

An early morning finds JN-4 trainers being prepared for the day's flying, with pilots checking their aircraft. Early models were underpowered with 90-horsepower Curtiss OX-5 engines, but later planes used 150-horsepower Hispano-Suiza engines, allowing a maximum speed of 79 miles per hour. Both pupil and instructor sat in open cockpits with dual controls, and the instructor communicated with the student in front over the noise of the motor through a speaking tube or by tapping him on the shoulder. (CA.)

These men are enjoying the end of a day of flying or repairing planes. Tents were necessary at Hazelhurst because barracks could not be built quickly enough to house the many soldiers stationed there during World War I; with wooden floors, they were adequate until winter. This snapshot shows much about soldiers' off-duty life. The newspapers, magazines, books, guitar, and a variety of household chairs, along with the sunshine and each other's company, imply that the men's smiles were not just for the photographer. (CA.)

If a buddy had a car, free time could also be spent going for a drive, probably to beaches in the summertime. Long Island Rail Road trains went into New York City, and there were service clubs operated by Red Cross, Knights of Columbus, and Salvation Army volunteers at nearby Mitchel Field. But Hazelhurst's hangout was this building, which provided a welcome alternative to mess hall chow. These enlisted men are enjoying this wartime afternoon together. (CA.)

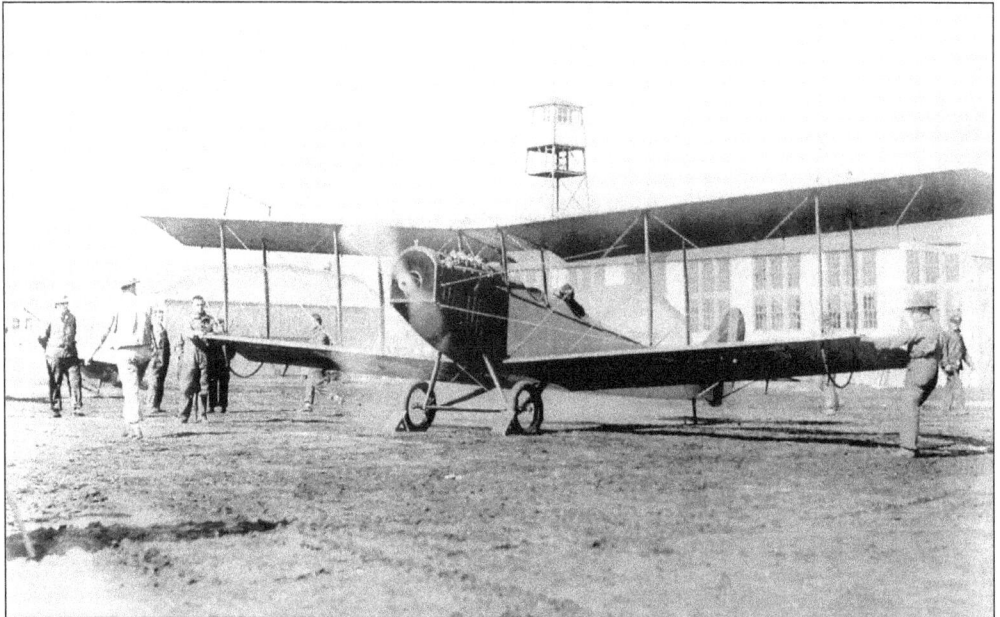

This Jenny is having its run-up to determine if the engine and instruments are working properly. Mechanics and pilots could tell much about an engine's performance by listening, and hopefully, any oil or water leaks should show before takeoff. The wheels are chocked, and two soldiers are holding the wingtips to steady the aircraft. The tower in the rear is not a control tower but was used to spot crashes and to dispatch fire trucks. The men with the wheelbarrow are filling the field's potholes. (CA.)

Before 1918, the army flew airmail cross-country in its flimsy trainers, and many crashes and deaths resulted from overly long flights in stormy weather and poor visibility. The Post Office Department then took over these flights, using mostly war surplus de Havilland DH-4 aircraft, which carried 400 pounds of mail. At Hazelhurst in 1924, a former army pilot has his parachute checked by his wife, a sure sign of her affection, since many of these post office planes also crashed. (CA.)

Landing at speeds of only 40 miles per hour, a pilot could expect to survive most crashes on his field; a mechanic would be sure of many days of work or a great many used spare parts after each bad landing. This JN-4 has nosed into Hazelhurst's grass field during World War I, flopping upside down. It was constructed of wooden frames over which cloth was stretched and then painted with dope to tighten it. Minor repairs could be made at any field with a mechanic on duty. (CA.)

Mitchel Field, created south of Hazelhurst Field in 1917, was named for a popular New York City mayor, John Purroy Mitchel, who died in an army aviation training accident. This 1933 aerial view shows its 10 hangars (which had checkered roofs to allow pilots to better identify the landing ground), along with many brick shop, warehouse, and cantonment buildings. The field became the major air base for the New York area; it now contains Nassau Community College and the Cradle of Aviation Museum. (CA.)

The 1st Aero Squadron, which saw service in the Mexican Expedition before it went overseas in World War I, flew de Havilland DH-4 reconnaissance planes. The unit's mission was to locate enemy troops and alert friendly forces by dropping a note to them that was tied to a rock. Early battery-operated radio sets soon saved the need to carry rocks. The pilot and observer are dressed for the 15,000-foot ceiling these planes attained. (CA.)

The Quartermaster Corps constructed permanent brick barracks, hangars, shops, warehouses, dining halls, a hospital, and officers quarters at Mitchel in the early 1930s. These Air Corps officer cadets are standing in company formations led by their senior cadet officers, much as is done at the Air Force Academy today. They will be marched to classes, meals, and chapel, and will obey orders given by upperclassmen as if they were commissioned officers. (CA.)

The probability of World War II brought the construction of temporary barracks and support buildings at Mitchel by 1941. Built by civilian contractors following standardized plans, they held some 120 airmen with double bunking. An unusual feature were the aqua medias, canopies extending above the windows to permit better ventilation during rainstorms, since troops could seldom return to close windows. This service road runs behind two rows of barracks; wider company streets are in front. (CA.)

One of Mitchel's hangars housed the 1930s tire shop, where these mechanics are remounting an aircraft tire on its rim. The sides of the hangars held shops and storerooms, with offices on the second level opening onto a balcony. The X-shaped truss kept the steel-framed building from swaying, and the sign on it reminds men that the typically male characteristic of not asking for assistance could lead to airplane accidents. Used as workshops, where oil and grease were common, hangars became grimy. (CA.)

Mitchel's paint shop was housed in a World War I wooden hangar; this civilian worker is painting our national insignia on a fabric wing. America's five-pointed star was replaced with a tri-color circle in 1918 because a white star looked like the German white-bordered cross in aerial combat. The circle had red on the outside, then blue, and a white center. The star was reauthorized after the war. (CA.)

With military spending limited, the army sought publicity and public support through various ploys. Long-distance and endurance flights were staged, Hollywood personalities were posed by planes, and yearly open houses were arranged. At Mitchel in the 1930s, 60 women in bathing suits and 25 smiling Air Corps officers are posing on the wings of a Boeing XB-15 long-range bomber. Even though the photograph was popular, this plane, the only one built, was a failure. (CA.)

This Mitchel parachute rigger is performing a periodic inspection of the silk Irwin Air Chutes used in 1940. Every 30 days, unused chutes were removed from their seats or backpacks and hung to be inspected for tears, damage from oil and chafing, and frayed shroud lines. When the chutes were repacked on the 45-foot-long table near the wall, bone packing sticks were used to tuck in the folds and smooth the wrinkles, and weighted shot bags held the canopy while shroud lines were straightened. (CA.)

Some Mitchel Field jobs were not cozy, as this meteorologist surely realizes. He is atop the operations building releasing a weather balloon in the 1940s. He will track the balloon with a theodolite to judge wind direction and speed at higher altitudes, and then he will telephone his observations to the weather office below. This procedure had to be done several times daily to ensure current information for pilots. A windsock, a long conical bag constructed to bend in two places depending on wind speed, indicated ground wind speed and direction. (CA.)

Mitchel Field volunteers ("You, you, you, and you!") are testing a self-inflating life raft for Air Corps bombers by paddling around in Long Island Sound off Glen Cove in the early 1940s. The rubber raft could hold a crew of 10 and was inflated by a carbon dioxide cylinder or a hand pump. Though the actual bomber crew would be in the water while their raft was inflating, these men are dry, but they still look concerned, possibly because their coxswain keeps them rowing. (CA.)

Life at army bases was made more pleasant by the establishment of post exchanges, general stores where enlisted men, officers, and their families could purchase sundry articles for their comfort and pleasure. Prices were lower than at civilian stores, as the post exchanges were not run for a profit. Mitchel's is shown in the 1940s; here, airmen could conveniently buy cigarettes, civilian clothing, extra uniforms, magazines, candy and snacks, and many other items. (CA.)

Mitchel's firehouse was the scene of non-firefighting activities in 1950; these men are repairing children's toys at Christmastime. Santa Claus, the stoutest of these men, will soon be riding on a fire engine around the base's housing areas, distributing donated and refurbished gifts to airmen's children. These men lived upstairs in the firehouse; when not responding to fires or crashes, they performed safety inspections in the base's many buildings, gave firefighting training, practiced, and maintained their equipment. (CA.)

There was nothing to do but wait for the crash truck after this 1917 accident. The Curtiss JN-4H advanced trainer's landing gear may have collapsed on the rough field, but the pilot seems quite unhurt. The many wires connecting the fabric-covered wings are bracings that are tightened by turnbuckles at their ends; they hold the wings together against the wooden struts and shape both wings symmetrically to stabilize the plane. Even with this damage, the plane will be repaired in Mitchel's shops. (CA.)

Every air base had its boneyard, and this is Mitchel's during World War I. Crashed planes that could not be repaired were stripped of major useable parts and dumped in this salvage yard. Mechanics needing a part that was not in stock would search through the piles of engines and fuselages hoping to find something they could use on another plane, as these two soldiers are doing. Because aircraft parts have to meet stringent airworthiness standards now, boneyards are no longer maintained. (CA.)

These Mitchel airmen will never have to be concerned about a crash landing. This postcard shows a popular way to send home a photograph during World War I, and the soldier with the X above his head (the man steering the plane) is likely the one who suggested to his buddy that they go into a penny arcade studio, where this gag photograph was taken. (LP.)

Several acres of farmland near Mitchel were fitted with fences and temporary barracks, and more than 1,000 Italian prisoners of war were housed there until Italy's 1943 surrender. Many worked on Long Island's farms, which was permitted because such work did not contribute directly to the war effort, and were paid for their labor. These men are loading their clothing bags and suitcases aboard a truck and will board the bus to begin their journey home. Some returned after repatriation and became American citizens. (CA.)

Roosevelt Field's many hangars and support buildings were taken over by both the navy and the army during World War II. In 1943, sailors are lined up in formation along the airfield side of Hangar E, one of six, together with a dog and one of the men's sons on the right. The navy operated schools and repair shops and provided long concrete runways where planes traveling between other bases could refuel and obtain maintenance. (CA.)

Army students are marching off in 1943 to the formerly civilian Roosevelt Aviation School, whose instructors in the white coveralls accompany them to the shops and hangars where their training as aviation mechanics will continue. Since there were already successful schools at Roosevelt Field, the military continued their operation, but with servicemen as their only students. The plane, whose engine has been removed, is used in the shops for students practicing their mechanical skills. (CA.)

Reading and resting were popular pastimes, and these World War II soldiers at Roosevelt Field are spending some of their off-duty hours catching up with the latest Hollywood gossip and, of course, cheesecake. Barracks were kept ready for inspection, with beds made, shoes polished and lined up, laundry bags secured to bedrails, and hanging clothes properly arranged in wall lockers. Pinups were seldom allowed in stateside barracks, though a girlfriend's photograph could be taped to the inside of a locker. (CA.)

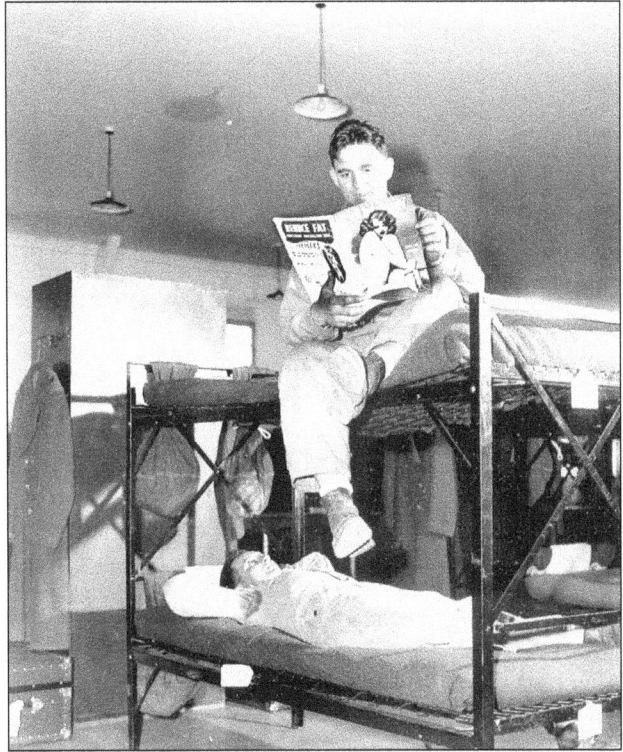

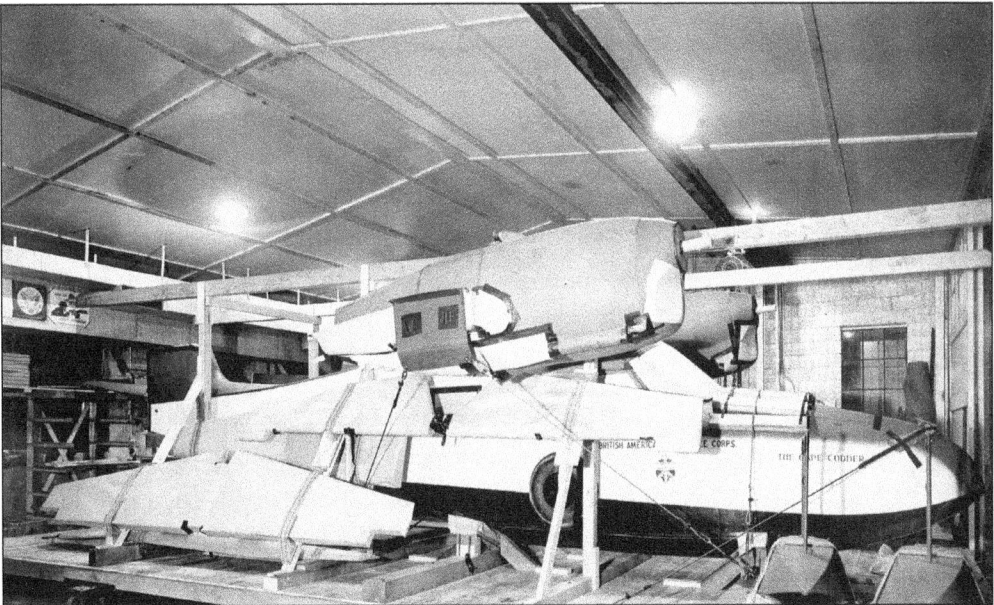

Dade Brothers played a specialized and essential role during World War II. Large airplanes could be ferried across the Atlantic to England, but fighters and other small aircraft did not have that range and had to be shipped, often as deck cargo. At Roosevelt Field, Dade packaged thousands of aircraft from plants around the country for overseas shipment. In one of Dade's buildings, a Grumman J4F Widgeon amphibian equipped as an air ambulance for Britain is ready for final crating. (CA.)

In 1917, Glenn Curtiss built the first complete aircraft research facility and factory on the Hempstead Plains just south of Hazlehurst Field. Its many buildings, seen in this 1920s aerial view, housed design departments, laboratories, machine shops, engine test stands, and a modern wind tunnel. Curtiss Jennies and OX engines were built here during World War I, and Curtiss purchased Hazelhurst in 1920 as a commercial airport and for his flying school. The main building survives; it is now used by an office supply company. (CA.)

The factory's main floor in 1928 houses the assembly line for Curtiss Model 53 Condor transport planes. Developed by Curtiss for the army as the B-2 bomber, the design was adapted for civilian purposes by enclosing the pilots' cockpit and adding 18 passenger seats. Two 625-horsepower Curtiss GV-1570 engines gave it a cruising speed of 125 miles per hour and a range of some 800 miles. Fuselage and tail assembly was done here, and the wings were attached later in a hangar. (CA.)

Six

AMERICA'S ARSENAL

In the fuselage department, Curtiss employees build the navy's first fighter designed to be launched from its first aircraft carrier, the *Langley*, a converted collier. The TS-1 was all wooden, both frame and skin, and was powered by a Lawrence 200-horsepower J-1 air-cooled radial engine. The wooden frames and aerodynamic surfaces were steam-bent, shaped, fitted, glued, and sanded in this room by skilled woodworkers who crafted the planes by hand; machine tools were only used to cut and drill the wooden members. (CA.)

Aircraft construction began in the drafting room, such as this one at Curtiss. Designers' sketches and ideas, along with the practical knowledge gained by experience and through wind-tunnel testing of scale models, were translated into working drawings. Draftsmen used India ink to draw the shape and specifications of the plane's parts, and of the dies and templates to manufacture them, on vellum. Sets of blueprints for the shops will next be made. (CA.)

No production plane was built without a wooden mockup being fabricated first, since many objects of fixed size—people, machine guns, seats, engines, and radios—had to be fitted together without snags. Visibility from the cockpit of a carrier plane, whose landing depended on the pilot's seeing the deck and the landing signal officer, is also best determined from a mockup. The four men inside Grumman's XJF-1 in 1932 are spending one of many days "flying" to see how everything fits. (NG.)

Grumman's first factory, a Baldwin garage, and its first airplane, the FF-1, are shown in 1931. The navy biplane was its first to have retractable landing gear. It carried two men: the pilot and an observer, who doubled as a gunner. With an aluminum frame and skin and a 700-horsepower Wright radial engine, it attained 200 miles per hour. The root of the lower wing is below the cockpit, which was enclosed by a canopy. Hanging above is an upper wing without its covering. (NG.)

By 1938, Grumman had moved from Baldwin to Curtiss Field in Valley Stream, then had moved again to Farmingdale, where a production line for navy F3F fighters was established in an old truck factory. This is the assembly area for the single-seat planes, which will soon join squadrons on the *Enterprise* and *Yorktown*. These were the last navy biplane fighters, as their speed was low (264 miles per hour) and their wings could not be folded when stowed on the carriers' hangar decks. (NG.)

The final assembly line of the F3F-2 was moved into Grumman's new Bethpage plant in 1938. The building was spacious compared with earlier facilities. During World War II, many additional buildings were constructed around the company's airfield. The jigs holding the planes' assemblies while they are being riveted are of wood. Soon, as mass-production techniques improved, metal fixtures were employed. (NG.)

Grumman used all available factory space on the island to manufacture carrier-based fighters, including the old Pan American Airways seaplane base in Port Washington. These 1944 employees are riveting wings held in metal jigs for F6F Hellcats, the navy's highly successful World War II combat aircraft. These will be trucked to Bethpage for attachment to fuselages completed there. Women worked alongside men in the shops and were valued for learning necessary mechanical skills and working long, fatiguing hours. (NG.)

Republic Aviation's P-47 Thunderbolt fighter was designed in 1940 by Alexander Kartveli and built at the company's Farmingdale plant and in Evansville, Indiana. The P-47D version won fame as a long-range fighter-escort for bombers over Germany and helped to lessen their losses to enemy fighters. In 1944, these workers are affixing the skin to a fuselage section using sheet holders, the small clamps sticking out of the aluminum. They keep the predrilled holes in alignment until rivets fasten the skin to frame members. (CA.)

The vastness of Republic's assembly building can be seen in this 1944 photograph, which shows less than a fourth of this plant's area. Adjoining factory buildings contained machine shops, foundries, subassembly areas, heat-treatment facilities, structural testing areas, and warehouses. P-47Ds for the American 8th Air Force are being assembled on the production line in the center, while fuselages stored along the side bear British Royal Air Force insignia. Others were produced for the Soviets and Free French, totaling more than 15,500 aircraft. (CA.)

Morale-building celebrations noted the completion of landmark-numbered planes. These Republic workers are pausing in 1947 after finishing the 100th forward fuselage assembly of an F-84 Thunderjet, the company's first jet-powered fighter. They saw aerial combat over North Korea in the 1950s and outfought the Soviet MIG-15. This fuselage section is attached to an overhead crane, which will carry it to another part of the plant to be attached to the rear section. (CA.)

This happy couple is not going out on a winter date. Two Republic research engineers are entering a cold chamber in 1943 to test canopies for the P-47. At the plane's 42,000-foot ceiling, plastics and metals could shatter or separate from the plane under flight stress. The room will be refrigerated to –65 degrees Fahrenheit, and the engineers will flex the canopy while results are filmed through an insulated window. (CA.)

Subcontractors made some assemblies used by airframe manufacturers. Liberty Aircraft Products in Farmingdale was one of the largest subcontractors. It was located in a factory built by Lawrence Sperry in 1918 and then occupied by Sherman Fairchild's aircraft company. Women are sewing fabric on the tail surfaces of an unidentified observation plane or primary trainer during World War II. A fabric covering, strengthened by doping, was suitable for use on control surfaces of slower aircraft, and its use helped conserve aluminum, which was needed for combat planes. (CA.)

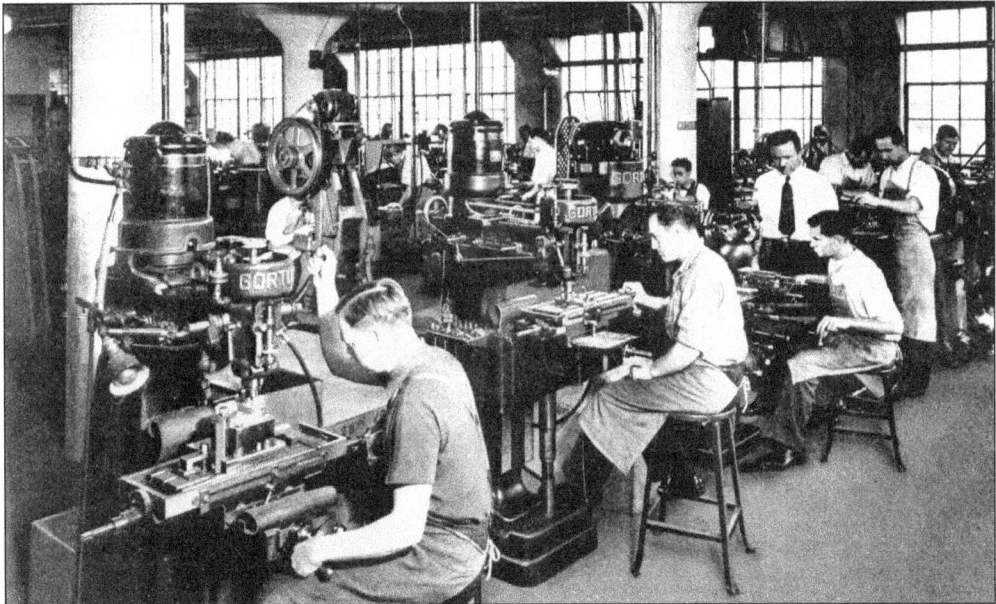

Because the manufacture of precision aircraft instruments is a specialized undertaking involving extensive research and the long-term training of skilled employees, airplane companies used quality flight instruments made by Bendix, Precision Instruments, Sperry, or Kollsman. This 1940s photograph shows a section of the machine shop of Kollsman's Elmhurst factory, where men are milling parts for altimeters or air speed indicators to very close tolerances. Great accuracy was necessitated by increases in the dive and climb speed of combat aircraft. (CA.)

Farmingdale's Ranger Engines operated test cells to run its engines before shipping. This is the control room of one cell in the 1930s. The instruments include a dynamotor scale indicating engine power, several manometers measuring air pressures or vacuums, tachometers, thermometers, and two stopwatches. The brick wall protected the operators if the engine exploded, and the throttles on the right allowed them to rev up the engine beyond specifications. Ranger engines were used in Fairchild PT-19 trainers and Grumman J4F-1 amphibians. (CA.)

Elmer Sperry, fascinated by toy gyroscopes as a boy, developed his interest into a company that made gyroscopic navigation devices, ship stabilizers, bombsights, aircraft machine-gun turrets, naval gun directors, and radar equipment. Sperry carbon-arc searchlights were made for the navy, and the Coast Artillery Corps used them to illuminate harbors. On the roof of the Sperry Gyroscope Company's Long Island City factory, several models await nightfall to be tested; the largest is an enclosed 60-inch searchlight for seacoast defense. (CA.)

This 1917 photograph illustrates the problem caused by constructing something larger than the freight elevator. The Sperry horse-drawn 36-inch searchlight wagon was built for the army for antiaircraft spotting; it will soon join its companion generator wagon on the Long Island City street. The connecting cable is on the rear reel; another cable allowed the operator to control the light's movements from a distance, as one could not see an airplane when looking up at the searchlight beam's glare. (CA.)

Sperry gyro-compasses were the company's best-loved product, found on all American naval and merchant ships. The compass repeater and helm in the wheelhouse was fondly nicknamed "Metal Mike" for its strength and reliability. Aboard the first nuclear submarine, *Nautilus*, in 1958, a Sperry engineer confers with a ship's officer at the master gyro-compass, used to navigate this submarine under the icecap at the North Pole. (CA.)

The Rockaway Naval Air Station was established in 1917 as a base for airships, observation balloons, and Curtiss flying boats, all of which were used to patrol against German submarines. Here, a Type C, twin-engine dirigible is leaving the hangar, while two unpowered balloons sag, weighted by sandbags. All were made lighter than air with hydrogen, which was dangerous but available. The balloons lifted an observer in a basket, who communicated with the ground by telephone; the telephone wires were part of the tether rope. (CA.)

Curtiss NC "Nancy" flying boats were designed to locate and fight enemy submarines, but the company's Garden City factory did not finish building and testing any before the war's end. One, though, became the first airplane to fly across the Atlantic. In 1919, three NCs took off from Rockaway on the first leg of their crossing, but only the NC-4 reached Portugal, after stopping in Nova Scotia, Newfoundland, and the Azores. This is the NC-1 being assembled at Rockaway in 1918. (CA.)

America's first submarine base was John Holland's factory in New Suffolk. His initial navy contract was for a steam-powered submersible that quenched its boiler fires before submerging (but still roasted the crew). His second sub, *Plunger*, is moored next to his *Shark*. Off Oyster Bay in 1905, Theodore Roosevelt became the first president to travel aboard a sub. He wrote that he "did not like to have the officers and enlisted men think I wanted them to try things I was reluctant to try myself." (NA.)

Navy observers at Castle Gould in Sands Point watch the flight of this pilotless, radio-controlled target drone in the 1950s. Daniel Guggenheim, always keenly interested in aviation, donated his waterfront estate to an aeronautical research institute, and by 1945, it had become the home of the Naval Training Device Center. Research and development of prototype mechanical and electronic equipment was conducted until the facility closed in 1967. The estate is now a Nassau County park. (CA.)

World War II lookout towers, where pairs of men and women worked two-hour shifts reporting aircraft activity, were located throughout the island. Some were enclosed, but volunteers here watch from an open tower. Powerful binoculars allow one observer to follow a plane's flight while the other telephones Mitchel Field's filter center, where a plotting board with moveable markers enables reports from many stations to be coordinated and a decision to be made about sounding air-raid sirens and dispatching fighter interceptors. (CA.)

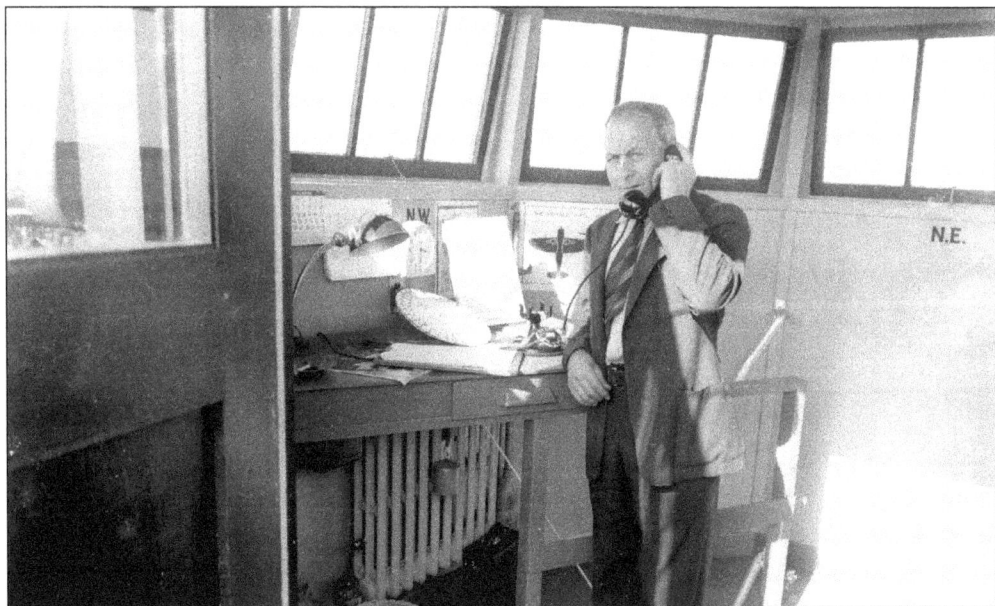

Civilian spotters telephoned their observations over regular telephone lines to the filter center, where they were plotted by military personnel. At Central Islip's station, built by the maintenance department of the state mental hospital atop its administration building, an employee is calling in to test the connection at the start of his shift. The aircraft recognition silhouette on the wall is of Republic's P-47, the most frequently spotted plane. (LI.)

Seven

THE HOME FRONT
AND AFTERWARDS

Civil defense activities prepared residents for possible attack and aided home-front morale by making civilians feel they were contributing to the war effort. At Central Islip State Hospital, employees are loading an air-raid "victim" into a makeshift ambulance, some wearing gas masks as the sign suggests. He will be taken to a temporary operating room set up in a nearby field. Nightly blackouts and frequent drills, which included schoolchildren huddling in corridors away from windows, were part of wartime life. (LI.)

Montauk's Fort Pond Bay was chosen as the site of a World War II naval torpedo test range. Shallow enough to permit the recovery of torpedoes that failed to surface after testing and close to the navy's Narragansett Bay development facility, the range operated until 1945. The nearby Montauk Manor, developer Carl Fisher's 1930s resort hotel, was used to house the range's naval officers. (LP.)

Coastguardsmen and their dogs patrolled Long Island's beaches during World War II, especially after eight German saboteurs landed from submarines at Amagansett and at Jacksonville, Florida, in 1942, carrying explosives to destroy war plants, particularly aluminum refineries and aircraft factories. The four men who landed at Amagansett were discovered by an unarmed coastguardsman walking beach patrol; when he returned with help, they had disappeared. One saboteur soon turned in the others; six were executed, and two were imprisoned and then deported after the war. (NA.)

German espionage agents sent coded messages about merchant ship sailings from New York's harbor using the civilian Telefunken radio station in Sayville. One message revealed the *Lusitania*'s departure information, enabling a submarine to torpedo the vessel. Other transmissions involved German minister Zimmermann's offer to Mexico to attack America in exchange for southwestern states. The navy seized the transmitter in 1915 and used the station thereafter; this 1990s photograph shows the abandoned transmitter halls and an FAA tower. (LP.)

Supposedly able to land a bomb in a pickle barrel from 20,000 feet, the Norden bombsight was secret, but a workman sold some of its secrets. The German spy ring he was a part of operated a clandestine radio station in Centerport; all 33 agents were arrested in 1941. The bombsight was not as accurate as it was publicized to be, but it was still a precise mechanical analog computer, and it was used successfully throughout the war. This Mitchel Field trainer moves over a scale map, guided by the bombardier and bombsight atop it. (CA.)

Many civilian facilities were taken over for military use in World War II, and Famous Players Studio in Astoria became the Signal Corps Pictorial Center. A training film is being shot in a huge sound studio, as in Hollywood, but the director and crew are soldiers here. In place of the Marx Brothers' comedy and Gloria Swanson's love scenes, these actors are depicting soldiers in a combat command post telephoning about some enemy action. (NA.)

Grumman employees are performing at a birthday party on April 20, 1943, and their lunch-hour audience is sorry that the honoree was not able to attend, as they would love to be able to say a few things to him. Popular patriotic songs such as "In der Fuehrer's Face" and "Remember Pearl Harbor" are sung in commemoration of what all hope will be Adolph Hitler's very last birthday. The TBFs and Hellcats they build will help that goal. (NG.)

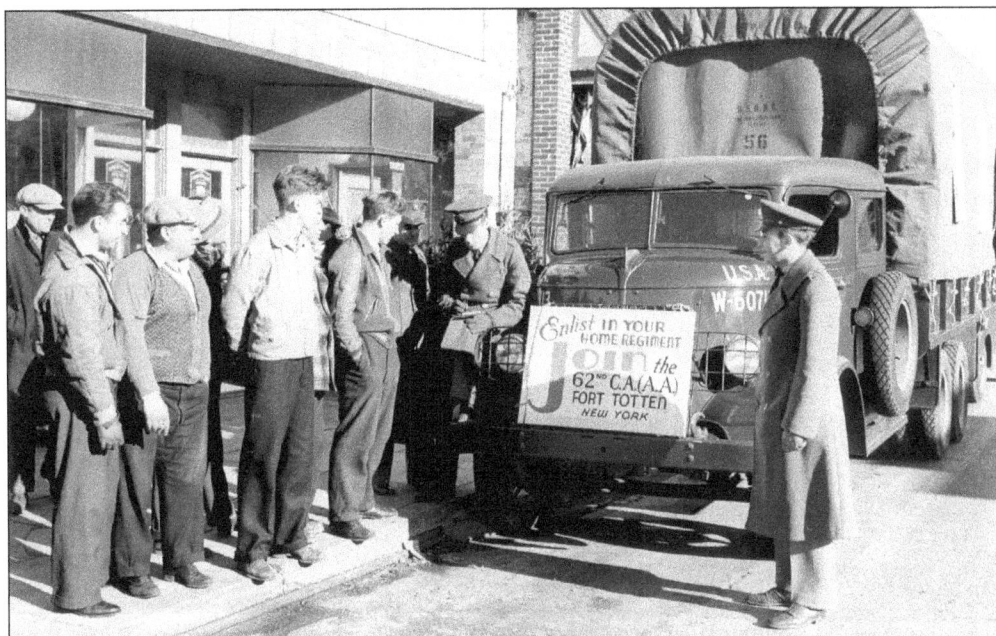

Recruiting began before America entered World War II, and these young men on a Queens street are being encouraged to join the 62nd Coast Artillery Regiment, stationed at nearby Fort Totten. This unit manned three-inch and .50-caliber antiaircraft batteries rather than coastal guns. Most of the members of this unit and similar antiaircraft regiments, and nearly all Coast Artillery personnel, were sent overseas later to defend against enemy aircraft attacks, which were then unlikely against the United States itself. (TM.)

Despite deferments for working in a crucial industry, these Grumman employees were drafted in 1943; here, they are being given a party and their final paychecks. By 1945, as seen in a later Grumman photograph, most employees were older men, as so many younger men were needed by the military to replace combat losses. This department manufactured fuel tanks, and the reminder that the machinery here was used to mill only aluminum and duralumin kept ferrous metals from contaminating the process. (NG.)

Every Long Island community placed an honor roll in a prominent location to honor its service personnel. Stars were affixed by names to indicate combat deaths. A small banner with a blue star for each living son or daughter serving was also placed in the front window of a home. A "Gold Star Mother" was one who had lost one of her children, and it was a title of honor. Here, Kings Park's classic memorial is unveiled downtown by scouts and local civilians. (KP.)

Wartime rationing involved food, footwear, rubber tires, and gasoline, but many Long Islanders needed their cars to get to work. Grumman organized car pools and helped employees obtain treasured B stickers and ration coupons for their cars. The company also took over several closed gas stations near the plant to enable employees to obtain the gas they needed during periodic shortages without having to drive around looking for a station that had some. (NG.)

Scrap drives were held to collect scarce materials necessary in war industries. Tinfoil from cigarette packages, abandoned cars and machinery, metal toothpaste tubes, newspapers, lard, and old tires were continuously collected, and there were special drives to collect pennies for their copper (1943 pennies contained none) and house keys for their brass (every hardware store soon had a barrelful as top bureau drawers were emptied). In Kings Park, residents piled scrap metal along the tracks for collection. (KP.)

Industrial scrap collection also yielded millions of tons of precious war materials, especially when they came from an aircraft factory such as Grumman's. Aluminum scrap was also obtained by collecting kitchen pots and pans, but trimmings of this metal and its alloys from plane production greatly added to America's raw materials and saved the time and cost of refining the aluminum. These men are weighing boxes of scrap, which will then be loaded aboard a train heading for a smelter. (NG.)

Most homes, businesses, and industries on Long Island were heated by coal, which was also needed by war plants and military camps. Shortages developed, abetted by a miners' strike, and lowered thermostats and rationing resulted. Grumman encouraged employees to take scrap wood from its salvage yard as kindling or stove wood or to supplement their emptying coal bins. Old packing cases and warehouse pallets fill the yard in 1943, and several men are spending their lunchtime gathering pieces. (NG.)

United Service Organization volunteers worked throughout World War II to raise the morale of America's fighting men and women. USO Clubs opened near camps, civilians spent their evenings talking and dancing with men who otherwise would find very little unpaid female company, and Broadway and Hollywood personalities—Bob Hope being one of the most famous—traveled around the world entertaining troops near the front lines. This 1944 photograph shows part of Hempstead's USO Club, with doughnuts and coffee aplenty. (LP.)

Grumman employees and their company yearly donated millions of items such as candy, razors, food, and soap to send to troops overseas at Christmastime, showing them that people back home had not forgotten them and making the season a little more joyous. Some of the boxes will be mailed to servicemen known to employees, perhaps as pen pals whom they have never met. The Red Cross will distribute the remainder to hospitalized men or to those who would receive no other gift. (NG.)

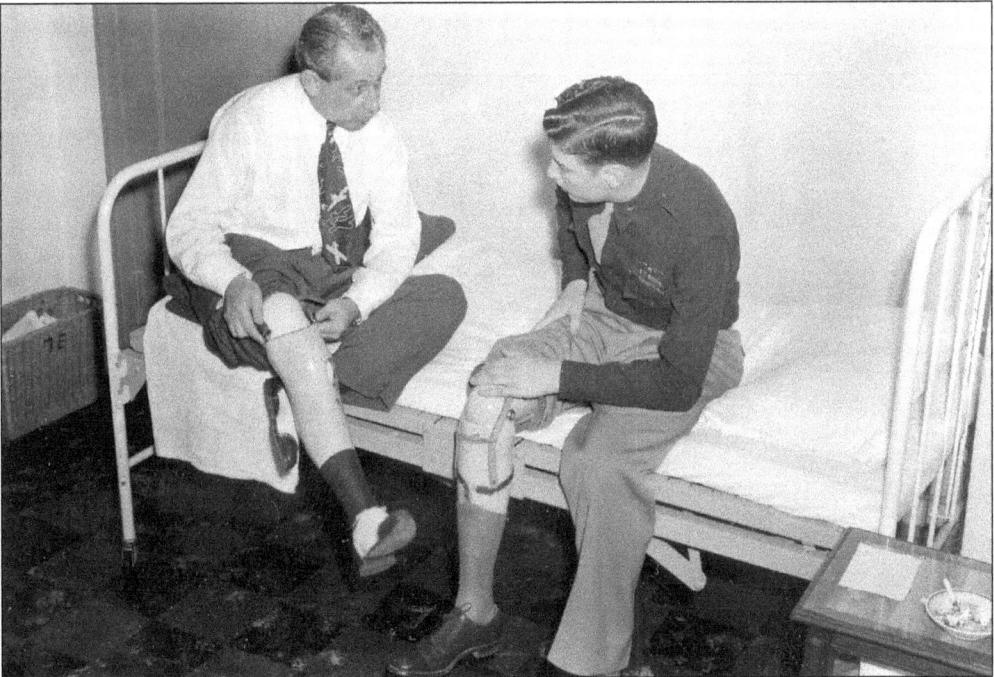

Many Long Islanders contributed to the war at home in their own unique ways. Aside from designing and testing aircraft in the 1930s that evolved into Republic's P-47, Maj. Alexander de Seversky, who lost a leg in aerial combat for Russia in World War I and continued combat and civilian flying with an artificial leg, helped during World War II by visiting servicemen who had also lost limbs. His example helped this soldier realize that he had a useful life ahead of him. (CA.)

Mail was one of the best morale builders for soldiers on the home front and overseas, and these Waves are processing V-mail from Long Islanders to sailors aboard ships and at bases around the world. The letters will be microfilmed and small negatives will be flown overseas in place of the bulky paper letters. Reproduced there, the letters will be delivered to their recipients. The Waves are reading them to ensure the addresses are valid and to screen some for a censor's approval. (LP.)

"Rosie the Riveter," the nickname give to women who volunteered in war industries, did not just work on the production floor. These 1940s employees are operating the control tower at Grumman's airfield, directing the takeoffs of completed aircraft being ferried to bases on three coasts to be loaded aboard aircraft carriers. Since every plane was also test flown, these ladies are arranging these flights and plane movements on the ground. VHF radio was used for ground-to-air communications, and blinker lamps were used for planes on the runway. (NG.)

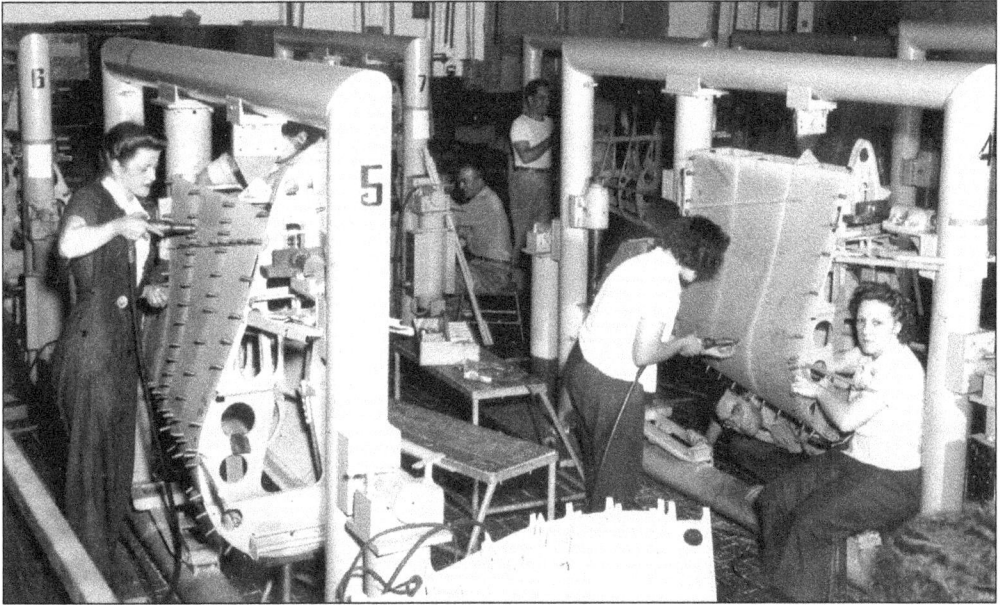

Grumman provided child-care facilities for its female employees, so important were their wartime skills to the navy's aircraft needs. These women are riveting fuselage subassemblies for F6F-5 Hellcat fighters. Clamped in rigid jigs and assembled to strict tolerances, all will be identical. Employees placed their coats on hangers to which a long pole was attached so the garments could be raised and hung high off the plant floor to save space. (NG.)

These Grumman production test pilots—one a navy pilot's wife, another a former Link trainer instructor, and the third a Civil Air Patrol pilot—are readying for a day's work by taking their parachutes and radio headsets from shelves in the pilot's room. After briefings, a check of the weather, and engine warm-up time, they will begin 20-minute test flights of the company's planes, taking them up for the first time to see if anything will go wrong. (NG.)

All wartime industries, including Grumman, met employee shortages by offering jobs to disabled people, assigning them to positions that they could fill with pride and adapting their facilities and machinery to accommodate their needs. This man is operating a hydraulic injection press whose controls have been lowered to a height that is comfortable for him. Disabled men and women often impressed their fellow workers with the quality and speed of their production work, and their efforts were crucial in meeting growing aircraft quotas. (NG.)

His foreman and a fellow Grummanite are encouraging this blind router department employee in 1943. He is filing the rough edges off an airplane part shaped by the spinning cutter of a routing machine and is testing the filed edges by touch. Dozens of blind employees, many accompanied by their guide dogs, were employed by Grumman in jobs in which their loss of sight did not prevent them from accomplishing much-needed work. (NG.)

War bond drives involving rallies, parades, and movie stars encouraged Americans to finance the war by purchasing government bonds, and children were asked to collect savings stamps that could later be exchanged for bonds. The money was spent on munitions and equipment. This 1942 Babylon parade features the wing of a destroyed Japanese Mitsubishi A6M "Zero" fighter, and the sign on the trailer connects bond purchases with downing enemy planes. (CA.)

Grumman employees are placing an F6F-5 Hellcat on the island in Times Square in 1945 to prepare for the Sixth War Loan drive. This navy fighter was specifically designed to combat the Japanese Zero fighter and was credited with destroying 5,250 enemy planes using its six .50-caliber wing-mounted machine guns. Some 12,275 Hellcats were produced. The Times Building is on the left, and a scaled-down replica of the Statue of Liberty is behind the plane. (NG.)

Grumman shop workers stand for a moment of silence on hearing of the death of Pres. Franklin D. Roosevelt on April 14, 1945, remembering all he had done to bring the country through difficult times in the Great Depression and World War II. Americans worldwide showed their respect and sorrow in similar ways. Hitler is said to have rejoiced, believing the president's death would somehow bring a German victory. After four years of military conscription, these remaining Grumman workers are mostly older men. (NG.)

Most of the island's military sites were abandoned after the war and returned to civilian uses. Since coast artillery forts were located in scenic areas, they became parks, with most fortifications left intact. Here, Fort Totten's Battery Graham, once mounting two 10-inch disappearing guns, has a lone visitor in 1996. The guns were removed in 1918, and all of the trees have grown since. This gun line and the uncompleted granite fort at the water's edge are part of New York City's park system. (GW.)

Charles Lindberg's smiling face looks out into Roosevelt's Hangar F in 1938, the result of painter Ailene Rhonie's talents. Her mural, on 120 feet of canvas, depicts Long Island's military and civilian aviation history from the earliest balloonists to Lindberg's historic 1927 cross-Atlantic solo flight, which began at Roosevelt Field. Most of the 500 faces, shown with planes, hangars, and airfields, were known to Rhonie, herself an aviator. The painting took three years to complete and is now being stored. (CA.)

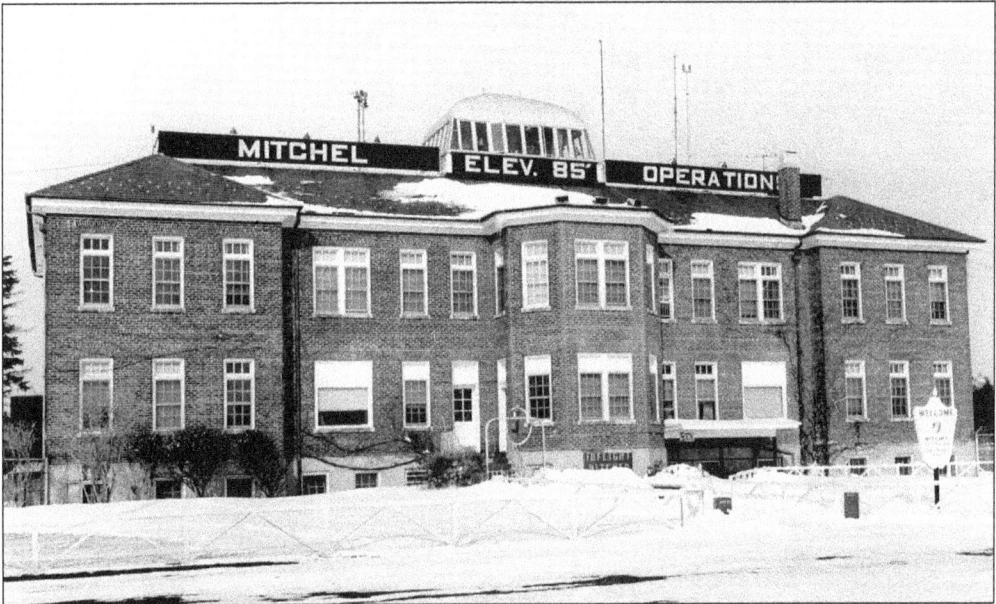

Mitchel Field's operations building, with its handsome control tower atop, is seen from the runway on a snowy day in the 1940s. When the field closed in 1961, Nassau Community College took over many of its buildings for classrooms and offices; the operations building was transformed into the student union, but with the tower removed. The elevation sign assisted pilots in setting altimeters and indicated the relative power needed for takeoff, as increased runway height means thinner air and less lift. (CA.)

ACKNOWLEDGMENTS

Many individuals and archives were especially helpful in the preparation of this history. Felice Ciccione, curator at Fort Wadsworth for the National Park Service's Gateway National Recreational Area, was as friendly and knowledgeable as she always has been. Joshua Stoff, curator at the Cradle of Aviation Museum at the old Mitchel Field, opened the museum's collection to us and was always very pleased to assist. Northrop-Grumman Corporation's History Center, staffed by retired employees, such as Larry Feliu, Mike Hlinko, Linn McDonald, and Bob Tallman, supplied invaluable photographs and information, as did Barry Moldano and Richard Cox, both of the Army Museum Service at Fort Hamilton. Jack Fine's large collection of Fort Totten photographs and those made available to us by such friends as Jonathan Prostak, Alex Holder, and Karl Schmidt also made this volume possible. And the Kings Park Heritage Museum's director, Leo Ostebo, supplied home-front snapshots that were of great interest.

Photographs are keyed as to source by these notations: (AH) Alex Holder collection; (CA) Cradle of Aviation Museum; (CM) Casemate Museum, Fort Monroe; (GN) Gateway National Recreational Area, Fort Wadsworth unit; (GW) Glen Williford collection; (HM) Fort Hamilton Army Museum; (JP) Jonathan Prostak collection; (KP) Kings Park Heritage Museum; (KS) Karl Schmidt postcard collection; (LC) Library of Congress; (LI) Long Island Studies Institute; (LP) Leo Polaski collection; (MH) Military History Institute, Carlisle Barracks; (NA) National Archives, College Park; (NG) Northrop-Grumman History Center, Bethpage; (NO) NOAA aerial photograph; (OM) Ordnance Museum, Aberdeen Proving Ground; (RD) Roger Davis postcard collection; (SC) Suffolk County Historical Society; (SH) Sandy Hook unit of Gateway National Recreational Area; (TM) Fort Totten Museum.

Base closings are nostalgic times for the military personnel involved, as most of them worked and lived in their buildings for years. In 1961, this air force captain, the field's operations officer, is closing and locking the control tower for the last time, the final flight having taken off that afternoon. Huge Xs will be painted at the ends of the runways, replacing their familiar compass designations and indicating that the field is permanently closed. (CA.)